T0327618

M-statistics

M-statistics

Optimal Statistical Inference
for a Small Sample

Eugene Demidenko

Dartmouth College
Hanover, NH, USA

Published by John Wiley & Sons, Inc., Hoboken, New Jersey.
Published simultaneously in Canada.

For general information on our other products and services or for technical support, please contact our Customer Care Department within the United States at (800) 762-2974, outside the United States at (317) 572-3993 or fax (317) 572-4002.

Wiley also publishes its books in a variety of electronic formats. Some content that appears in print may not be available in electronic formats. For more information about Wiley products, visit our web site at www.wiley.com.

Library of Congress Cataloging-in-Publication Data
Name: Demidenko, Eugene, 1948- author.
Title: M-statistics : optimal statistical inference for a small sample / Eugene Demidenko.
Description: Hoboken, NJ : Wiley, 2023. | Includes bibliographical references and index.
Identifiers: LCCN 2023000520 (print) | LCCN 2023000521 (ebook) | ISBN 9781119891796 (hardback) | ISBN 9781119891802 (adobe pdf) | ISBN 9781119891819 (epub)
Subjects: LCSH: Mathematical statistics.
Classification: LCC QA276.A2 D46 2023 (print) | LCC QA276.A2 (ebook) | DDC 519.5/4–dc23/eng20230410
LC record available at https://lccn.loc.gov/2023000520
LC ebook record available at https://lccn.loc.gov/2023000521

Cover Design: Wiley
Cover Image: © 500px Asia/Getty Images; Equations by Eugene Demidenko

Set in 11/13.5pt Computer Modern by Straive, Chennai, India

To my family

Contents

Preface

Mean and variance are the pillars of classic statistics and statistical inference. These metrics are appropriate for symmetric distributions but are routinely used when distributions are not symmetric. Moreover, the mean is not coherent with human perception of centrality – we never sum but instead look for a typical value, the mode, in statistics language. Simply put, the mean is a suitable metric for computers, but the mode is for people. It is difficult to explain why the mean and median are parts of any statistical package but not the mode. Mean and variance penetrate the theory of statistical inference: the minimum variance unbiased estimator is the landmark of classic statistics. Unbiased estimators rarely exist outside of linear statistical models, but even when they do, they may produce incomprehensible values of the squared correlation coefficient or variance component. This work offers a new statistical theory for small samples with an emphasis on the exact optimal tests and unequal-tail confidence intervals (CI) using the cumulative distribution function (cdf) of a statistic as a pivotal quantity.

The organization of the book is as follows. The first chapter outlines the limitations of classic statistics and uses the normal variance statistical inference for a brief introduction to our theory. It contains a section on some extensions of the Neyman-Pearson lemma to be used in subsequent chapters. The second and third chapters introduce two competing approaches: maximum concentration and mode statistics with optimal confidence intervals and tests. An optimal confidence interval (CI) with coverage probability going to zero gives birth to a maximum concentration or mode estimator depending on the choice of the CI. Two tracks for statistical inference are offered combined under one umbrella of M-statistics: maximum concentration (MC) theory stems from the density level test, short-length CI and implied MC parameter estimator. Mode (MO) theory includes the test, unbiased CI, and implied parameter MO estimator. Unlike the classic approach, we suggest different CI optimality criteria and respective estimators depending on the parameter domain. For example, the CI with the minimum length on the log scale yields a new estimator of the binomial probability and Poisson rate, both positive, even when the number of successes is zero. New criteria for efficient estimation are developed as substitutes for the variance and the mean square error. Chapter 4 discusses definitions of the p-value for asymmetric distributions – a generally overlooked but an important statistical problem. M-statistics is demonstrated in action in Chapter 5. Novel exact optimal CIs and statistical tests are developed for major statistical parameters: effect size, binomial probability, Poisson rate, variance component in the meta-analysis model, correlation coefficient, squared correlation coefficient, and coefficient of determination in linear model. M-statistics is extended to the multidimensional parameter in Chapter 6. The exact confidence regions and unbiased tests for normal mean and variance, the shape parameters of the beta distribution, and nonlinear regression illustrate the theory.

The R codes can be freely downloaded from my website

    www.eugened.org

stored at GitHub. By default, each code is saved in the folder C:\Projects\Mode\ via dump command every time it is called. If you are using a Mac or do not want to save in this directory, remove or comment (#) this line before running the code (sometimes external R packages must be downloaded and installed beforehand).

I would like to hear comments, suggestions, and opinions from readers. Please e-mail me at eugened@dartmouth.edu.

Dartmouth College   *Eugene Demidenko*
Hanover, NH, USA
April 2023

Chapter 1

Limitations of classic statistics and motivation

In this chapter, we discuss the limitations of classic statistics that build on the concepts of the mean and variance. We argue that the mean and variance are appropriate measures of the center and the scatter of *symmetric* distributions. Many distributions we deal with are *asymmetric*, including distributions of positive data. The mean not only has a weak practical appeal but also may create theoretical trouble in the form of unbiased estimation – the existence of an unbiased estimator is more an exception than the rule.

Optimal statistical inference for normal variance in the form of minimum length or unbiased CI was developed more than 50 years ago and has been forgotten. This example serves as a motivation for our theory. Many central concepts, such as unbiased tests, mode, and maximum concentration estimators for normal variance serve as prototypes for the general theory to be deployed in subsequent chapters.

The Neyman-Pearson lemma is a fundamental statistical result that proves maximum power among all tests with fixed type I error. In this chapter, we prove two results, as an extension of this lemma, to be later used for demonstrating some optimal properties of M-statistics such as the superiority of the sufficient statistic and minimum volume of the density level test.

M-statistics: Optimal Statistical Inference for a Small Sample, First Edition. Eugene Demidenko.
© 2023 John Wiley & Sons, Inc. Published 2023 by John Wiley & Sons, Inc.

1.1 Limitations of classic statistics

1.1.1 Mean

A long time ago, several prominent statisticians pointed out to limitations of the mean as a measure of central tendency or short *center* (Deming 1964; Tukey 1977). Starting from introductory statistics textbooks the mean is often criticized because it is not robust to outliers. We argue that the mean's limitations are *conceptually* serious compared to other centers, the median and the mode.

For example, when characterizing the distribution of English letters the mean is not applicable, but the mode is "e.". Occasionally, statistics textbooks discuss the difference between mean, mode, and median from the application standpoint. For example, consider the distribution of house prices on the real estate market in a particular town. For a town clerk, the most appropriate measure of the center is the mean because the total property taxes received by the town are proportional to the sum of house values and therefore the mean. For a potential buyer, who compares prices between small nearby towns, the most appropriate center is the mode as the typical house price. A person who can afford a house at the median price knows that they can afford 50% of the houses on the market.

Remarkably, modern statistical packages, like R, compute the mean and median as `mean(Y)` and `median(Y)`, but not the mode, although it requires just two lines of code

$$\mathtt{densY = density(Y)}$$

$$\mathtt{modeY = densY\$x[densY\$y == max(densY\$y)]}$$

where `Y` is the array of data. The centerpiece of the mode computation is the `density` function, which by default assumes the Gaussian kernel and the bandwidth computed by Silverman's "rule of thumb" (1986).

Consider another example of reporting the summary statistics for U.S. hourly wages (the data are obtained from the Bureau of Labor Statistics at https://www.bls.gov/mwe). Figure 1.1 depicts the Gaussian kernel density of hourly wages for 234,986 employees. The mean is almost twice as large as the mode because of the heavy right tail. What center should be used when reporting the *average* wage? The answer depends on how the center is used. The mean may be informative to the U.S. government because the sum of wages is proportional to consumer buying power and collected income taxes. The median has a clear interpretation: 50% of workers earn less than $17.10 per hour. The mode offers a better interpretation of the individual level as the typical wage – the point of maximum concentration of wages. In parentheses, we report the proportion

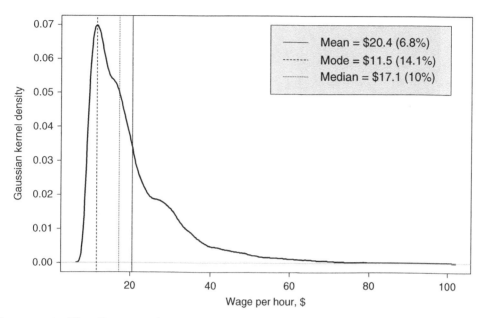

Figure 1.1: *The Gaussian kernel density for a sample of 234,986 hourly wages in the country. The percent value in the parentheses estimates the probability that the wage is within ±$1 of the respective center.*

of workers who earn ±$1 within each center. The mode has maximum data concentration probability – that is why we call the mode typical value. The mean ($20.40) may be happily reported by the government, but $11.50 is what people typically earn.

Mean is a convenient quantity for computers, but humans never count and sum – they judge and compare samples based on the typical value.

Figure 1.2 illustrates this statement. It depicts a NASA comet image downloaded from https://solarviews.com/cap/comet/cometneat.htm. The bull's-eye of the comet is the mode where the concentration of masses is maximum. Mean does not have a particular interpretation.

Mean is for computers, and mode is for people. People immediately identify the mode as the maximum concentration of the distribution, but we never sum the data in our head and divide it by the number of points – this is what computers do. This picture points out the heart of this book: the mean is easy to compute because it requires arithmetic operations suitable to computers. The mode requires more sophisticated techniques such as density estimation – unavailable at the time when statistics was born. Estimation of the mode is absent even in comprehensive modern statistics books. The time has come to reconsider and rewrite statistical theory.

Figure 1.2: *Image of comet C/2001 Q4 (NEAT) taken at the WIYN 0.9-meter telescope at Kitt Peak National Observatory near Tucson, Arizona, on May 7, 2004. Source: NASA.*

1.1.2 Unbiasedness

The mean dominates not only statistical applications but also statistical theory in the form of an unbiased estimator. Finding a new unbiased estimator is regarded as one of the most rewarding works of a statistician. However, unbiasedness has serious limitations:

1. The existence of unbiased estimator is an exception, not the rule. Chen (2003) writes "... the condition on unbiasedness is generally a strong one." Unbiased estimators mostly exist in the framework of linear statistical models and yet classic statistical inference like the Cramér-Rao lower bound or Lehmann-Sheffe theorem relies on unbiasedness (Rao 1973; Zacks 1971; Casella and Berger 1990; Cox and Hinkley 1996; Lehmann and Casella 1998; Bickel and Docksom 2001; Lehmann and Romano 2005). Unbiased estimators do not exist for simple nonlinear quantities such as the coefficient of variation or the ratio of regression coefficients (Fieller 1932).

2. Unbiasedness is not invariant to nonlinear transformation. For example, the square root of the sample variance is a positively biased estimator of the standard deviation. If an unbiased estimator exists for a parameter, it rarely exists for its reciprocal.

3. An unbiased estimator of a positive parameter may take a negative value, especially with small degrees of freedom. The most notorious examples are the unbiased estimation of variance components and the squared correlation coefficient. For example, as shown by Ghosh (1996), the unbiased nonnegative estimator of the variance component does not exist.

4. Variance and mean square error (MSE), as established criteria of statistical efficiency, suffer from the same illness: they are appropriate measures of scattering for symmetric distributions and may not exist even in simple statistical problems.

Note that while we criticize the unbiased estimators, there is nothing wrong with unbiased statistical tests and CIs – although the same term unbiasedness is used, these concepts are not related. Our theory embraces unbiased tests and CIs and derives the mode (MO) and maximum concentration (MC) estimator as the limit point of the unbiased and minimum length CI, respectively, when the coverage probability approaches zero.

1.1.3 Limitations of equal-tail statistical inference

Classic statistics uses the equal-tail approach for statistical hypothesis testing and CIs. This approach works for symmetric distributions or large sample sizes. It was convenient in the pre-computer era when tables at the end of statistics textbooks were used. The unequal approach, embraced in the present work, requires computer algorithms and implies optimal statistical inference for any sample size. Why use a suboptimal equal-tail approach when a better one exists? True, for a fairly large sample size, the difference is negligible but when the number of observations is small say, from 5 to 10 we may gain up to 20% improvement measured as the length of the CI or the power of the test.

1.2 The rationale for a new statistical theory

The classic statistical inference was developed almost 100 years ago. It tends to offer simple procedures, often relying on the precomputed table of distributions printed at the end of books. This explains why until now equal-tail tests and CIs have been widely used even though for asymmetric distributions the respective inference is suboptimal. Certainly, for a moderate sample size, the difference is usually negligible but when the sample size is small the difference can be considerable. Classic equal-tail statistical inference is outdated. Yes, unequal-tail inferences do not have a closed-form solution, but this should not serve as an excuse for using suboptimal inference. Iterative numeric algorithms are an inherent part of M-statistics – Newton's algorithm is very effective starting from classic equal-tail values.

In classic statistics, CIs and estimators are conceptually disconnected. In M-statistics, they are on the same string of theory – the maximum concentration (MC) and mode (MO) estimators are derived as the limit points of the CI when the confidence level approaches zero.

Finally, we want to comment on the style of this work. Novel ideas are illustrated with statistical inference for classic distribution parameters along with numeric implementation and algorithms that are readily available for applications.

Our theory of M-statistics proceeds on two parallel competing tracks:

- MC-statistics: density level test, short CI, and implied MC estimator.

- MO-statistics: unbiased test, unbiased CI, and implied MO estimator.

In symbolic form,
$$\mathbf{M} = \mathbf{MC} + \mathbf{MO}.$$

Our new tests and CIs are exact and target small samples. Neither the expected value of the estimator nor its variance/mean square error plays a role in our theory. In classic statistics, the CI and the estimator are not related conceptually. In M-statistics, the latter is derived from the former when the confidence level approaches zero.

Our theory relies on the existence of a statistic S with the cumulative distribution function (cdf) $F(s; \theta)$. Since $F(S; \theta)$ has the uniform distribution on (0,1), this cdf can be viewed as a pivotal quantity. Statistical inference based on pivotal quantity is well known but limited mainly because it exists in rare cases. On the other hand, derivation of the cdf of any statistic is often feasible via integration.

The organization of the book is as follows: We start with a motivating example of the CI for the normal variance dating back to classic work by Neyman and Pearson (1933). These CIs serve as a template for our further development of unbiased CIs and tests in the general case. Not surprisingly, sufficient statistics produce optimal CIs. Classic fundamental concepts such as the Neyman-Pearson lemma and Fisher information receive a new look in M-statistics. We argue that for a positive parameter, the length of the CI should be measured on a relative scale, such as the logarithmic scale leading to a new CI and implied MC estimator. This concept applies to binomial probability and the Poisson rate. The cdf is just a special case of a pivotal quantity for constructing unbiased CIs and MC estimators, which are illustrated with a simple meta-analysis variance components model. This example emphasizes the optimal properties of the MC estimator of the variance, which is in the center of the maximum concentration probability and is always positive. The correlation coefficient, squared correlation coefficient, and coefficient of determination in the linear model with fixed predictors are the next subjects of application of M-statistics. Although these fundamental statistics parameters were introduced long ago until recently exact unbiased CIs and statistical tests were unknown. Our work is finalized with the extension of M-statistics to the multidimensional parameter. Novel exact confidence regions and tests are derived and illustrated with two examples: (a) simultaneous testing of the normal mean and standard deviation, and (b) non-linear regression.

Finally, the goal of the book is not to treat every single statistical inference problem and apply every method for CI or hypothesis testing – instead, the goal is to demonstrate a new theory. Much work must be done to apply the new statistical methodology to other practically important statistical problems and develop efficient numerical algorithms.

1.3 Motivating example: normal variance

This section reviews the exact statistical inference for the variance of the normal distribution: shortly, normal variance. First, we discuss CI, and then we move on to statistical hypothesis testing. Our point of criticism is the widespread usage of the equal-tail CI and hypothesis test that silently assume a symmetric distribution. For example, the equal-tail statistical inference for the normal variance is used in a recent R package `DescTools`. However, the distribution of the underlying statistic, that is, the sum of squares is not symmetric and has a chi-square distribution. We argue that equal-tail CIs were convenient in the pre-computer era when tables were used. Instead, we popularize a computationally demanding unequal-tail CI: specifically, the unbiased CI and the CI with the minimum expected length (the short CI). Although unequal-tail CIs have been known for many years they are rarely applied and implemented in modern statistical software.

Here we apply unequal-tail statistical inference to normal variance as a motivating example and an introduction to our M-statistics theory. This example will serve as a benchmark for many novel methods of statistical inference developed in this book.

1.3.1 Confidence interval for the normal variance

To motivate our M-statistics, we revive the classic confidence interval (CI) for the normal variance, σ^2, having n independent normally distributed observations

$Y_i \sim \mathcal{N}(\mu, \sigma^2)$, $i = 1, 2, ..., n$. The basis for the CI is the fact that $\sigma^{-2}S$, where $S = \sum_{i=1}^{n}(Y_i - \overline{Y})^2$, has a chi-square distribution with $n - 1$ degrees of freedom (df).

Traditionally to construct a double-sided CI for normal variance (σ_L^2, σ_U^2), the equal-tail CI is used where the limits are computed by solving the equations $C_{n-1}(S/\sigma_L^2) = 1 - \alpha/2$ and $C_{n-1}(S/\sigma_U^2) = \alpha/2$ that yield

$$C_{n-1}(S/\sigma_L^2) - C_{n-1}(S/\sigma_U^2) = \lambda, \qquad (1.1)$$

where C_{n-1} denotes the chi-square distribution with $n - 1$ degrees of freedom (df) and $\lambda = 1 - \alpha$ is the desired coverage probability. If $q_{\alpha/2} = C_{n-1}^{-1}(\alpha/2)$ and $q_{1-\alpha/2} = C_{n-1}^{-1}(1 - \alpha/2)$ are the quantiles of the chi-square distribution the lower and upper confidence limits are $\sigma_L^2 = S/q_{1-\alpha/2}$ and $\sigma_L^2 = S/q_{\alpha/2}$. This interval covers the true σ^2 with the exact probability λ. The equal-tail CI is offered in various textbooks, handbooks, and statistical software, such as SAS, STATA, and R, despite being biased and not having a minimal expected length.

The short unequal-tail CI was introduced almost 100 years ago by Neyman and Pearson (1936). Anderson and Bancroft (1952) and Tate and Klett (1959) applied this CI to the normal variance; El-Bassiouni (1994) carried out comparisons with other CIs. Two asymmetric CIs for σ^2 are revived next: the minimum expected length and the unbiased CI. The former is a representative of the maximum concentration (MC) and the latter is a representative of the mode (MO) statistics.

Short CI

To achieve exact coverage of σ^2 the quantiles do not necessarily have to be chosen to have equal-tail probabilities. An obvious choice, given the observed S, is to find σ_L^2 and σ_U^2 such that they satisfy (1.1) and $\sigma_U^2 - \sigma_L^2$ is minimum. Since minimization of $\sigma_U^2 - \sigma_L^2$ is equivalent to minimization of $\sigma_U^2/S - \sigma_L^2/S$ upon change of variables $q_L = S/\sigma_U^2$ and $q_U = S/\sigma_L^2$, we replace (1.1) with

$$C_{n-1}(q_U) - C_{n-1}(q_L) = \lambda \qquad (1.2)$$

and arrive at the minimization of $1/q_L - 1/q_U$ as

$$\min_{C_{n-1}(q_U) - C_{n-1}(q_L) = \lambda} (1/q_L - 1/q_U). \qquad (1.3)$$

This problem is solved by the Lagrange multiplier technique with the function defined as

$$\mathcal{L}(q_L, q_U, \nu) = 1/q_L - 1/q_U - \nu[C_{n-1}(q_U) - C_{n-1}(q_L) - \lambda],$$

where $\nu > 0$ is the Lagrange multiplier. Differentiating $\mathcal{L}$ with respect to the unknown quantiles, we obtain the necessary conditions for the minimum:

$$\frac{\partial \mathcal{L}}{\partial q_L} = -1/q_L^2 + \nu c_{n-1}(q_L) = 0, \quad \frac{\partial \mathcal{L}}{\partial q_U} = 1/q_U^2 - \nu c_{n-1}(q_U) = 0,$$

where c_{n-1} denotes the probability density function (pdf) of the chi-square distribution with $n-1$ df:

$$c_{n-1}(s) = \frac{1}{2^{(n-1)/2}\Gamma((n-1)/2)} s^{(n-3)/2} e^{-s/2}, \quad s \geq 0. \tag{1.4}$$

After eliminating the Lagrange multiplier, we arrive at the following equation for q_L and q_U:

$$q_U^2 c_{n-1}(q_U) - q_L^2 c_{n-1}(q_L) = 0 \tag{1.5}$$

or equivalently,

$$c_{n+3}(q_U) - c_{n+3}(q_L) = 0 \tag{1.6}$$

Hence, the optimal quantiles that produce the minimal expected length CI are found by solving equations (1.2) and (1.5). After q_L and q_U are determined, the short CI (σ_L^2, σ_U^2) has the limits $\sigma_L^2 = S/q_U$ and $\sigma_U^2 = S/q_L$. Shao (2003, Theorem 7.3) generalized the short-length CI to any unimodal distribution having a pivotal quantity of a special kind.

Now we turn our attention to solving equations (1.2) and (1.5) using Newton's algorithm (Ortega and Rheinboldt 2000) starting from the equal-tail quantiles. Using the density function (1.4) one can show that equation (1.5) simplifies to

$$(n+1)(\ln q_U - \ln q_L) + (q_L - q_U) = 0. \tag{1.7}$$

The system of equations (1.2) and (1.7) is solved iteratively for (q_L, q_U) as

$$\begin{bmatrix} q_L \\ q_U \end{bmatrix}_{k+1} = \begin{bmatrix} q_L \\ q_U \end{bmatrix}_k - \begin{bmatrix} \delta_L \\ \delta_U \end{bmatrix}_k, \quad k = 0,1,2,..., \tag{1.8}$$

where k is the iteration index and the 2×1 adjustment vector is given by (iteration index not shown)

$$\begin{bmatrix} \delta_L \\ \delta_U \end{bmatrix} = \begin{bmatrix} -c_{n-1}(q_L) & c_{n-1}(q_U) \\ 1 - (n+1)/q_L & -1 + (n+1)/q_U \end{bmatrix}^{-1}$$

$$\times \begin{bmatrix} C_{n-1}(q_U) - C_{n-1}(q_L) - \lambda \\ (n+1)\ln(q_U/q_L) + (q_L - q_U) \end{bmatrix}. \tag{1.9}$$

We start iterations from the equal-tail quantiles: $q_{L0} = q_{\alpha/2}$ and $q_{U0} = q_{1-\alpha/2}$. Our practice shows that it requires only three or four iterations to converge. Note that the optimal quantiles are not random and depend on the confidence level,

λ, and n, but not on statistic S. After q_L and q_U are computed the short CI for normal variance takes the form $(\sigma_L^2, \sigma_U^2) = (S/q_U, S/q_L)$. The R function that implements Newton's iterations is var.ql with the call

$$\texttt{var.ql(n,adj,alpha,eps = 1e-05,maxit = 100)}$$

In this function, n is the sample size (n), adj is the adjustment parameter for n in the form n-adj as the coefficient at $\ln q_U - \ln q_L$, alpha passes the value of α, eps is the tolerance difference between iterations, and maxit is the maximum number of Newton's iterations. The function returns a two-dimensional vector at the final iteration, (q_L, q_U). An example of the call and output is shown here.

```
> var.ql(n=20,adj=-1,alpha=0.05)
[1]  9.899095 38.327069
```

If S is the observed sum of squares with $n = 20$ the short 95% CI for σ^2 is $(S/38.327069, S/9.899095)$. We use the adj argument to accommodate computation of other quantiles for normal variance, such as for unbiased CIs and tests (see the following section).

The limits of the equal- and unequal-tail 95% CI are compared in Figure 1.3 as functions of n. Since the limits of the CI have the form S/q, the reciprocal of quantiles, $1/q$, are shown on the y-axis. The plot at left depicts $1/q_{0.975}$ and $1/q_L$, and the plot at right depicts $1/q_{0.025}$ and $1/q_U$, where q_L and q_U are solutions of (1.2) and (1.7) found by Newton's iterations (1.9). The limits of the short CI are shifted downward. Figure 1.4 depicts the percent length reduction of the short CI compared with the traditional equal-tail CI. Not surprisingly, the difference disappears with n, but for small sample sizes, such as $n = 5$, the unequal-tail CI is shorter by about 30%.

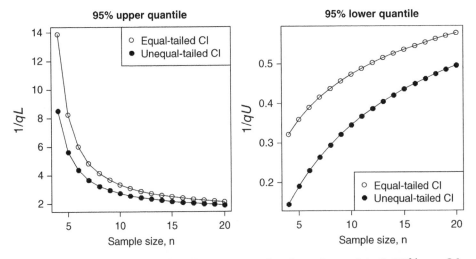

Figure 1.3: *Comparison of the short unequal-tail and equal-tail 95% confidence limits for the normal variance using reciprocal quantiles.*

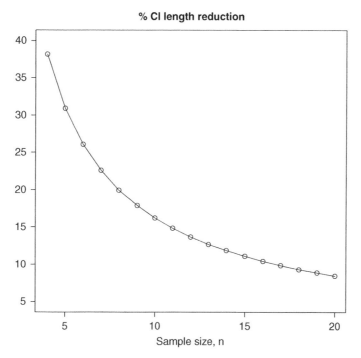

Figure 1.4: *Percent length reduction of the short CI compared to the traditional equal-tailed CI for normal variance.*

Unbiased CI

The traditional equal-tail and short-length CIs are biased for asymmetric distributions, such as the chi-square, meaning the coverage probability of the "wrong" variance, σ_*^2, may be greater than the coverage probability of the true variance $\sigma^2 \neq \sigma_*^2$. In this section, we choose the quantiles q_L and q_U that make the CI unbiased. The coverage probability of σ_*^2 is

$$\Pr(\sigma_L^2 \leq \sigma_*^2 \leq \sigma_U^2) = \Pr(\sigma_L^2/S \leq S/\sigma^2 \leq \sigma_U^2/S)$$
$$= \Pr(\sigma_*^2 \sigma^{-2} q_L \leq \sigma^{-2} S \leq \sigma_*^2 \sigma^{-2} q_U)$$
$$= C_{n-1}(\sigma_*^2 \sigma^{-2} q_U) - C_{n-1}(\sigma_*^2 \sigma^{-2} q_L).$$

We demand that this probability reaches its maximum at $\sigma_*^2 = \sigma^2$, which requires the necessary condition

$$\frac{d(C_{n-1}(\sigma_*^2 \sigma^{-2} q_U) - C_{n-1}(\sigma_*^2 \sigma^{-2} q_L))}{d\sigma_*^2}\bigg|_{\sigma_*^2 = \sigma^2}$$
$$= \sigma^{-2}(q_U c_{n-1}(q_U) - q_L c_{n-1}(q_L)) = 0.$$

Thus we conclude that to make the CI $(S/q_L, S/q_U)$ unbiased the following must hold:

$$q_U c_{n-1}(q_U) - q_L c_{n-1}(q_L) = 0. \tag{1.10}$$

After some trivial algebra, we arrive at a similar system of equations where (1.7) is replaced with

$$(n-1)(\ln q_U - \ln q_L) - (q_U - q_L) = 0. \tag{1.11}$$

Again Newton's algorithm (1.8) applies with $n+1$ in (1.9) replaced with $n-1$. More specifically, the delta-vector takes the form

$$\begin{bmatrix} \delta_L \\ \delta_U \end{bmatrix} = \begin{bmatrix} -c_{n-1}(q_U) & c_{n-1}(q_L) \\ 1 - (n-1)/q_L & (n-1)/q_U - 1 \end{bmatrix}^{-1}$$
$$\times \begin{bmatrix} C_{n-1}(q_U) - C_{n-1}(q_L) - \lambda \\ (n-1)\ln(q_U/q_L) - (q_U - q_L) \end{bmatrix}.$$

The quantiles are computed by the same R function `var.ql` with `adj=1`. The call `var.ql(n=20,adj=1)` returns `9.267006 33.920798`. If S is the sum of squares with $n = 20$, the 95% unbiased CI for σ^2 is $(S/33.920798, S/9.267006)$. Since

$$(9.267006^{-1} - 33.920798^{-1}) - (9.899095^{-1} - 38.327069^{-1}) = 0.0035012 > 0$$

the length of the unbiased CI is greater than the length of the short CI, as it is supposed to be.

1.3.2   Hypothesis testing for the variance

The silent presumption of symmetry and implied equal-tail statistical inference penetrates the current practice of hypothesis testing as well. Although several authors have noted that the two-sided equal-tail test is biased, it is predominantly used in major statistical textbooks and popular software packages. In this section, we illustrate how to derive an unbiased test using unequal-tail probabilities. To avoid confusion, we emphasize that the meaning of "unbiased" test is different from the "unbiased" estimator criticized earlier.

The two-sided variance test for $H_0 : \sigma^2 = \sigma_0^2$ versus $H_A : \sigma^2 \neq \sigma_0^2$ with (non-random) q_L and q_U satisfying the condition in equation (1.2) rejects the null if the observed value $\sigma_0^{-2}S$ falls outside of the interval (q_L, q_U). To derive the power of the test we find the probability that $\sigma_0^{-2}S$, where $\sigma^{-2}S$ has a chi-square distribution with $n-1$ df and $\sigma^2 \neq \sigma_0^2$, falls between q_L and q_U:

$$\Pr(q_L \leq \sigma_0^{-2}S \leq q_U) = \Pr(S \leq \sigma_0^2 q_U) - \Pr(S \leq \sigma_0^2 q_U)$$
$$= C_{n-1}(q_U \sigma_0^2/\sigma^2) - C_{n-1}(q_L \sigma_0^2/\sigma^2).$$

The power function of the test dual to the CI is the complementary probability,

$$P(\sigma^2; \sigma_0^2) = 1 + C_{n-1}(q_L \sigma_0^2/\sigma^2) - C_{n-1}(q_U \sigma_0^2/\sigma^2). \qquad (1.12)$$

By evaluating the derivative at $\sigma^2 = \sigma_0^2$, it is easy to see that the equal-tail test is biased. However, elementary calculus proves that $dP/\sigma^2 = 0$ at $\sigma^2 = \sigma_0^2$ if q_L and q_U satisfy the system (1.11), the same as for the unbiased CI. Thus we conclude that the unbiased CI produces a dual unbiased test – this is a general result that has been known for a long time (Rao 1973, p. 470). The density level test will be introduced in Section 2.3 and the test for variance is discussed in Example 2.14.

Figure 1.5 depicts the power function of four tests for the null hypothesis $\sigma^2 = \sigma_0^2 = 1$ with $n = 20$ and $\alpha = 0.05$. The equal-tail test is biased because the tangent line of the power function (1.12) at $\sigma^2 = 1$ has a positive slope. The tangent line of the unbiased test with q_L and q_U derived from (1.11) is parallel

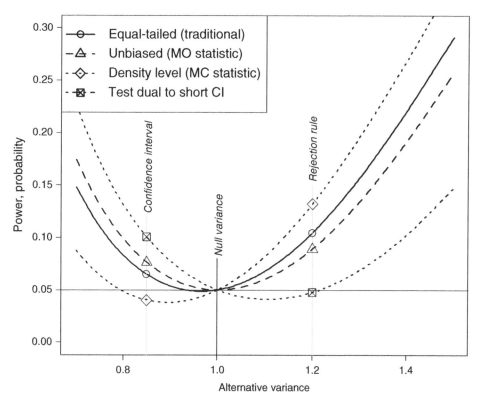

Figure 1.5: *The power functions of three tests for the two-sided variance hypothesis $H_0 : \sigma^2 = 1$ with $n = 20$ and $\alpha = 0.05$. Symbols at the alternatives $\sigma^2 = 0.85$ and $\sigma^2 = 1.2$ indicate the simulation-derived power values through the respective confidence interval and rejection rules. The density level test is discussed in Example 2.14. This graph was created with the R function* **vartest**.

to the x-axis. The power function at $\sigma^2 = 1.2$ is checked through simulations (number of simulations = 10 millions) as the proportion of simulations for which S/σ_0^2 is outside of the acceptance interval (q_L, q_U). The test dual to short CI is discussed in Section 4.3, and Example 4.18 illustrates its application to normal variance. All the tests are exact, but just for the unbiased test, the power function does not go below the significance level $\alpha = 0.05$ (the equal-tail test is just slightly biased).

1.3.3   MC and MO estimators of the variance

We define an estimator as the limit point of the CI when the confidence level goes to zero, $\lambda \to 0$. For normal variance the estimator of σ^2 takes the form S/q where $q = \lim_{\lambda \to 0} q_L = \lim_{\lambda \to 0} q_U$. Different CIs produce different MC estimators. When the minimum expected length CI is used, as follows from (1.5) or (1.6), the limit point is found from equation $c'_{n+3}(q) = 0$, the mode of the chi-square distribution with $n + 3$ df.

Although the following fact is well-known, we prove it for completeness of the presentation.

Proposition 1.1 *The mode of the chi-square distribution with n df is $n - 2$.*

Proof. The pdf of the chi-square distribution with n df is given by

$$c_n(s) = \frac{1}{2^{n/2}\Gamma(n/2)} s^{n/2-1} e^{-s/2}, \quad s \geq 0.$$

To simplify the derivation, take the derivative of the log-transformed pdf and set it to zero:

$$\frac{d\ln c_n}{ds} = \left(\frac{n}{2} - 1\right)\frac{1}{s} - \frac{1}{2} = 0.$$

Solving for s yields the maximizer of c_n, mode = $n - 2$. □

Since the mode of c_{n+3} is $n + 1$, the minimum expected length CI yields the following MC estimator of the variance:

$$\widehat{\sigma}_{MC}^2 = \frac{1}{n+1}\sum_{i=1}^{n}(Y_i - \overline{Y})^2. \tag{1.13}$$

When the confidence level of the unbiased CI approaches zero, we arrive at the MO estimator

$$\widehat{\sigma}_{MO}^2 = \frac{1}{n-1}\sum_{i=1}^{n}(Y_i - \overline{Y})^2, \tag{1.14}$$

the traditional unbiased estimator of the normal variance. Generally, it's not true that the unbiased CI yields an unbiased estimator. As we shall see from

Section 2.5.1, this estimator can be viewed as the MC estimator on the log scale.

When $\lambda \to 0$ the equal-tail CI converges to

$$\widehat{\sigma}_{ET}^2 = \frac{1}{q_{0.5}} \sum_{i=1}^{n} (Y_i - \overline{Y})^2,$$

where $q_{0.5}$ is the median of the chi-square distribution with $n - 1$ df. For large n, we have $q_{0.5} \simeq n - 1$, which is in agreement with the general fact that for large n, the difference between equal- and unequal-tail CIs vanishes.

The ideas and concepts of this section will give birth to M-statistics applied to the general case laid out in the following chapters.

1.3.4 Sample size determination for variance

Power calculation and sample size determination are frequent tasks when planning statistical studies. They are especially important in biomedical and pharmaceutical applications. In general terms, sample size determination solves the problem of finding the sample size, n, that yields the required power (p) given the type I error (α), and the null and alternative parameter values. In this section, we discuss computational issues related to finding n when testing the normal variance.

We aim to find n such that $P(\sigma^2; \sigma_0^2) = p$ where P is defined by (1.12). All parameters must be defined before computation of the sample size: σ_0^2, σ^2, $\alpha = 1 - \lambda$, and power p. Popular values are $p = 0.8$ and $\alpha = 0.05$. An approximate sample size can be derived from the generic formula based on the large-sample normal approximation (Demidenko 2007):

$$n = \frac{(\Phi^{-1}(1 - \alpha/2) + \Phi^{-1}(p))^2}{\delta^2} V,$$

where δ is the difference between the alternative and null values and V is the n times asymptotic variance of the underlying statistic. Since the variance of $\widehat{\sigma}^2$ is approximately $2\sigma^4/n$, we arrive at the following rough estimate for the required sample size:

$$n = \frac{(\Phi^{-1}(1 - \alpha/2) + \Phi^{-1}(p))^2}{(\sigma^2 - \sigma_0^2)^2} 2\sigma^4. \tag{1.15}$$

The exact sample size to reach power p can be obtained graphically by plotting the power function (1.12) as a function of n in some range around the approximate n with the power value covering the required p, and finding n for which the deviation from p is minimum. See Example 1.2 and the R code sampVAR.

The exact n can be found numerically, treating n as continuous, with a modified Newton's algorithm (Ortega and Rheinboldt 2000). The original Newton's algorithm requires differentiation of $P(\sigma^2; \sigma_0^2)$ with respect to n, a formidable task. Instead, we employ the large n approximation

$$P(\sigma^2; \sigma_0^2) \simeq \Phi \left(\sqrt{n} \frac{\sigma^2 - \sigma_0^2}{\sqrt{2}\sigma^2} \mp \Phi^{-1}(1 - \alpha/2) \right), \qquad (1.16)$$

where $-$ is used when $\sigma^2 > \sigma_0^2$ and $+$ otherwise. From this formula, we derive the approximation of the derivative as

$$\frac{d}{dn} P(\sigma^2; \sigma_0^2) \simeq \phi \left(\sqrt{n} \frac{\sigma^2 - \sigma_0^2}{\sqrt{2}\sigma^2} \mp \Phi^{-1}(1 - \alpha/2) \right) \frac{\sigma^2 - \sigma_0^2}{2\sqrt{2n}\sigma^2}. \qquad (1.17)$$

Hence the modified Newton's algorithm takes the form

$$n_{k+1} = n_k - \frac{P_k - p}{d_k}, \qquad (1.18)$$

where P_k is the exact power (1.12) evaluated at $n = n_k$ and d_k is the right hand side of (1.17). Typically, it takes three or four iterations to converge starting from approximate n defined by (1.15).

This method, with some modifications, can be applied to other sample size determinations, as described later in the book. The following example illustrates the sample size determination for stock volatility.

Example 1.2 *Stock volatility is an important risk factor to consider when playing the stock market. The null hypothesis is that the volatility of a stock measured as the variance of the log-transformed stock price on the percent scale is 0.2, the standard volatility, versus the alternative that the variance is 0.3 for a high-gain-high-risk stock. Assuming that the log-transformed observations follow a normal distribution, how many observations does one need to reject the null when it holds with probability 5% and accept the alternative when it holds with probability 80%?*

The power function for testing $H_0 : \sigma_0^2 = 0.2$ versus the alternative $H_A : \sigma^2 = 0.3$ is given by (1.12) with q_L and q_U found from (1.2). We need to solve the equation $P(\sigma^2; \sigma_0^2) = 0.8$ for n using the iterations (1.18). Figure 1.6 illustrates the power functions and the required n. The R code sampVAR computes the sample size and creates the figure. The solid line depicts the exact power (1.12) and the dotted line depicts a large-sample size approximation (1.16). A negative sign is used because $\sigma^2 > \sigma_0^2$. We note that the difference between the approximated and exact n is substantial, almost 50%.

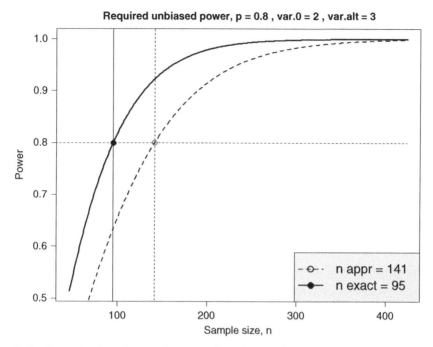

Figure 1.6: *Sample size determination for the stock prices in Example 1.2. The exact and approximated sample sizes are quite different.*

1.4 Neyman-Pearson lemma and its extensions

1.4.1 Introduction

Studying optimal properties of M-statistics requires some general preliminary results that are tightly connected to the celebrated Neyman-Pearson lemma (Neyman and Pearson 1933). These results will be instrumental for studying minimum-length CIs and minimum-volume confidence bands, optimal properties of density level tests, sufficient statistics, and extension of M-statistics to the multidimensional parameter. Before formulating the Neyman-Person lemma we set up the stage by describing the problem of the optimal test when only two hypotheses are involved.

The problem setup for statistical testing of the null hypothesis $H_0 : \theta = \theta_0$ versus the alternative $H_A : \theta \neq \theta_1$ is as follows. Let $\mathbf{Y}$ be an n-dimensional observed vector with the probability density function (pdf) under the null as $f_0(\mathbf{y}) = f(\mathbf{y}; \theta_0)$ and as $f_1(\mathbf{y}) = f(\mathbf{y}; \theta_1)$ under the alternative. The goal is to find the test of size $0 < \alpha < 1$ that maximizes the power: that is,

$$\int_A f_1(\mathbf{y}) d\mathbf{y}, \tag{1.19}$$

where $A \subset R^n$ is the rejection region. That is, if $\mathbf{Y} \in A$ then the null hypothesis is rejected. The fact that the test with rejection region A has type I error α is expressed as

$$\int_A f_0(\mathbf{y}) d\mathbf{y} = \alpha. \tag{1.20}$$

The problem of the optimal test reduces to finding set A that maximizes the power, or symbolically,

$$\max_A \int_A f_1(\mathbf{y}) d\mathbf{y}$$

under the restriction (1.20). Neyman and Pearson solved this problem by suggesting the rejection region be

$$A = \left\{ \mathbf{y} : \frac{f_1(\mathbf{y})}{f_0(\mathbf{y})} \geq c \right\}, \tag{1.21}$$

where $c > 0$ is such that

$$\int_{f_1(\mathbf{y})/f_0(\mathbf{y}) \geq c} f_0(\mathbf{y}) d\mathbf{y} = \alpha.$$

Test (1.21) is often referred to as the likelihood ratio test (LRT).

Example 1.3 *Application of the LRT to testing of the normal mean with unit variance.*

To illustrate the LRT, we consider testing the null hypothesis $H_0 : \mu = \mu_0$ against $H_1 : \mu = \mu_1$, where the alternative mean μ_1 is fixed, having a random sample $Y_1, Y_2, ..., Y_n$ with independent and identically normally distributed (iid) observations with mean μ and unit variance. Since the joint pdf is

$$f(y_1, y_2, ..., y_n; \mu) = (2\pi)^{-n/2} e^{-\frac{1}{2} \sum_{i=1}^n (y_i - \mu)^2},$$

where $\mu = \mu_0$ or $\mu = \mu_1$, according to the LRT test the null hypothesis, is rejected when

$$-\frac{1}{2} \sum_{i=1}^n (Y_i - \mu_1)^2 + \frac{1}{2} \sum_{i=1}^n (Y_i - \mu_0)^2 \geq c,$$

where c is to be found from equation (1.20). The rejection equation simplifies after some algebra:

$$-\frac{1}{2} \sum_{i=1}^n (Y_i - \mu_1)^2 + \frac{1}{2} \sum_{i=1}^n (Y_i - \mu_0)^2 = n(\mu_1 - \mu_0)\overline{Y} - \frac{n}{2}(\mu_1^2 - \mu_0^2),$$

where $\overline{Y}$ is the average. Consider two cases. First, assume that $\mu_1 > \mu_0$. Then the condition in equation (1.20) is equivalent to

$$\Pr\left(n(\mu_1 - \mu_0)(Z/n + \mu_0) - \frac{n}{2}(\mu_1^2 - \mu_0^2) \geq c\right) = \alpha.$$

where $Z = \sqrt{n}(\overline{Y} - \mu_0)$ is the standard normal variable. Thus instead of c, it is easier to find c' such that $\Pr(Z \geq c') = 1 - \Phi(c') = \alpha$, which gives $c' = \Phi^{-1}(1 - \alpha)$. Therefore for $\mu_1 > \mu_0$, the LRT rejects the null if $\sqrt{n}(\overline{Y} - \mu_0) > \Phi^{-1}(1 - \alpha)$. Second, assume that $\mu_1 < \mu_0$. Then c' is found from $\Pr(Z < c') = \Phi(c') = \alpha$, which gives $c' = \Phi^{-1}(\alpha)$. Finally since $\Phi^{-1}(\alpha) = -\Phi^{-1}(1 - \alpha)$, we deduce that the rejection region for the most efficient LRT test is

$$\begin{cases} \sqrt{n}(\overline{Y} - \mu_0) > \Phi^{-1}(1 - \alpha) \text{ if } \mu_1 > \mu_0 \\ \sqrt{n}(\overline{Y} - \mu_0) < -\Phi^{-1}(1 - \alpha) \text{ if } \mu_1 < \mu_0 \end{cases} \tag{1.22}$$

We make an important comment on the test. The reader may ask why the rejection area is not $\sqrt{n}|\overline{Y} - \mu_0| > \Phi^{-1}(1 - \alpha/2)$, which is more familiar. The reason is that in the Neyman & Pearson paradigm, the alternative is fixed at value μ_1, but in the Fisherian paradigm, is not. Consequently, in the former paradigm, the rejection rule is (1.22) and depends on μ_1, but in the latter paradigm the rule is $\sqrt{n}|\overline{Y} - \mu_0| > \Phi^{-1}(1 - \alpha/2)$ and does not depend on μ_1. More discussion on the difference between the two paradigms and the respective tests is found in Section 6.1.

1.4.2 Two lemmas

The following formulation of the Neyman-Pearson lemma will be used. The proof of the lemma is found in many statistics textbooks, such as Casella and Berger (1990), Schervish (2012), Shao (2003), and Lehmann and Romano (2005), among others. However, for the sake of completeness the proof, following the proof given in Hogg et al. (2006), is provided next.

Lemma 1.4 *Neyman-Pearson. Let $p(\mathbf{y})$ and $q(\mathbf{y})$ be two positive continuous pdfs defined on $\mathbf{y} \in R^n$, and A be a set in R^n such that*

$$\int_{p(\mathbf{y})/q(\mathbf{y}) \leq k} p(\mathbf{y}) dy = \int_A p(\mathbf{y}) dy, \tag{1.23}$$

for some $k > 0$. Then

$$\int_{p(\mathbf{y})/q(\mathbf{y}) \leq k} q(\mathbf{y}) dy \geq \int_A q(\mathbf{y}) dy. \tag{1.24}$$

Proof. To shorten the notation define the set

$$B = \{\mathbf{y} : p(\mathbf{y})/q(\mathbf{y}) \le k\}.$$

To prove (1.24) we may exclude the common part $A \cap B$ from the domain of integration. Using notation A^c for a set complementary to A we rewrite the inequalities (1.23) and (1.24) as

$$\int_{B \cap A^c} p(\mathbf{y}) d\mathbf{y} = \int_{A \cap B^c} p(\mathbf{y}) d\mathbf{y}, \quad \int_{B \cap A^c} q(\mathbf{y}) d\mathbf{y} \ge \int_{A \cap B^c} q(\mathbf{y}) d\mathbf{y}. \tag{1.25}$$

Since on set B, we have $q(\mathbf{y}) \ge k^{-1} p(\mathbf{y})$, we obtain

$$\int_{B \cap A^c} q(\mathbf{y}) d\mathbf{y} \ge \frac{1}{k} \int_{B \cap A^c} p(\mathbf{y}) d\mathbf{y} = \frac{1}{k} \int_{A \cap B^c} p(\mathbf{y}) d\mathbf{y}.$$

But

$$B^c = \{\mathbf{y} : p(\mathbf{y})/q(\mathbf{y}) > k\} = \{\mathbf{y} : p(\mathbf{y}) > kq(\mathbf{y})\},$$

and therefore

$$\frac{1}{k} \int_{A \cap B^c} p(\mathbf{y}) d\mathbf{y} \ge \frac{1}{k} \int_{A \cap B^c} kq(\mathbf{y}) d\mathbf{y} = \int_{A \cap B^c} q(\mathbf{y}) d\mathbf{y},$$

which proves the second inequality of (1.25), and consequently (1.24).   □

Note that in the traditional formulation, p and q are referred to as the pdf at the null and alternative hypotheses, respectively. The sets A and $\{\mathbf{y} : p(\mathbf{y})/q(\mathbf{y}) \le k\}$ are referred to as the rejection regions for a general test and the likelihood-ratio test, respectively. Equation (1.23) means the two tests have the same size and the inequality (1.24) means the likelihood ratio test is no less powerful.

Now we derive the following integral inequality to be used for proving optimal properties of M-statistics, including the superiority of sufficient statistics.

Lemma 1.5 *Let $p(\mathbf{y})$ be a positive continuous pdf, and $W(\mathbf{y})$ and $Q(\mathbf{y})$ be continuous functions of $\mathbf{y} \in R^n$ such that $E(Q) = \int_{R^n} Q(\mathbf{y})p(\mathbf{y})d\mathbf{y}$ exists. Then*

$$\int_{Q(\mathbf{y}) \le s_1} p(\mathbf{y}) d\mathbf{y} = \int_{W(\mathbf{y}) \le s_2} p(\mathbf{y}) d\mathbf{y} \tag{1.26}$$

implies

$$\int_{Q(\mathbf{y}) \le s_1} Q(\mathbf{y})p(\mathbf{y}) d\mathbf{y} \le \int_{W(\mathbf{y}) \le s_2} Q(\mathbf{y})p(\mathbf{y}) d\mathbf{y} \tag{1.27}$$

for any given s_1 and s_2.

Proof. The proof involves three cases: function Q is nonnegative, possibly negative but bounded from below, and unbounded.

When $Q(\mathbf{y}) \geq 0$, without loss of generality we can assume that Q is not a zero function because then the lemma holds, and therefore $E(Q) > 0$. Consequently, we may assume that $s_1 > 0$. We start by proving a slightly different result:

$$\int_{Q(\mathbf{y}) \geq s_1} p(\mathbf{y})d\mathbf{y} = \int_{W(\mathbf{y}) \geq s_2} p(\mathbf{y})d\mathbf{y} \qquad (1.28)$$

implies

$$\int_{Q(\mathbf{y}) \geq s_1} Q(\mathbf{y})p(\mathbf{y})d\mathbf{y} \geq \int_{W(\mathbf{y}) \geq s_2} Q(\mathbf{y})p(\mathbf{y})d\mathbf{y}. \qquad (1.29)$$

Indeed, introduce a pdf $q(\mathbf{y}) = Q(\mathbf{y})p(\mathbf{y})/E(Q)$ and the set $A = \{W(\mathbf{y}) \geq s_2\}$. Define k such that $\{p(\mathbf{y})/q(\mathbf{y}) \leq k\} = \{Q(\mathbf{y}) \geq s_1\}$ with an obvious choice $k = s_1^{-1} \int Q(\mathbf{y})p(\mathbf{y})d\mathbf{y}$. We employ the Neyman-Pearson lemma which leads to the desired inequality (1.29) under the condition (1.28). Now we work on replacing $\geq$ with $\leq$ to arrive at (1.27) under condition (1.26) through complementary quantities:

$$\int_{Q(\mathbf{y}) \leq s_1} p(\mathbf{y})d\mathbf{y} = 1 - \int_{Q(\mathbf{y}) \geq s_1} p(\mathbf{y})d\mathbf{y},$$

$$\int_{Q(\mathbf{y}) \leq s_1} Q(\mathbf{y})p(\mathbf{y})d\mathbf{y} = E(Q) - \int_{Q(\mathbf{y}) \geq s_1} Q(\mathbf{y})p(\mathbf{y})d\mathbf{y},$$

and the same for integrals that involve $W(\mathbf{y}) \geq s_2$.

Now consider the case when Q satisfies (1.26) but may be negative and bounded from below, $Q(\mathbf{y}) > Q_0$. Introduce a positive function $\widetilde{Q}(\mathbf{y}) = Q(\mathbf{y}) - Q_0 > 0$ with $0 < E(\widetilde{Q}) < \infty$. Then (1.26) can be written as

$$\int_{\widetilde{Q}(\mathbf{y}) \leq s_1 - Q_0} p(\mathbf{y})d\mathbf{y} = \int_{W(\mathbf{y}) \leq s_2} p(\mathbf{y})d\mathbf{y} \qquad (1.30)$$

which, due to the previous proof, implies

$$\int_{\widetilde{Q}(\mathbf{y}) \leq s_1 - Q_0} \widetilde{Q}(\mathbf{y})p(\mathbf{y})d\mathbf{y} \leq \int_{W(\mathbf{y}) \leq s_2} \widetilde{Q}(\mathbf{y})p(\mathbf{y})d\mathbf{y}.$$

But

$$\int_{\widetilde{Q}(\mathbf{y}) \leq s_1 - Q_0} \widetilde{Q}(\mathbf{y})p(\mathbf{y})d\mathbf{y} = \int_{\widetilde{Q}(\mathbf{y}) \leq s_1 - Q_0} Q(\mathbf{y})p(\mathbf{y})d\mathbf{y} - Q_0 \int_{\widetilde{Q}(\mathbf{y}) \leq s_1 - Q_0} p(\mathbf{y})d\mathbf{y},$$

$$\int_{W(\mathbf{y}) \leq s_2} \widetilde{Q}(\mathbf{y})p(\mathbf{y})d\mathbf{y} = \int_{W(\mathbf{y}) \leq s_2} Q(\mathbf{y})p(\mathbf{y})d\mathbf{y} - Q_0 \int_{W(\mathbf{y}) \leq s_2} p(\mathbf{y})d\mathbf{y}.$$

The second terms in both equations do not affect the inequality because

$$\int_{\widetilde{Q}(\mathbf{y}) \le s_1 - Q_0} p(\mathbf{y}) d\mathbf{y} = \int_{W(\mathbf{y}) \le s_2} p(\mathbf{y}) d\mathbf{y}$$

due to (1.26). Since $\widetilde{Q} > 0$ we refer to the previous case and therefore condition (1.30) implies

$$\int_{\widetilde{Q}(\mathbf{y}) \le s_1 - Q_0} Q(\mathbf{y}) p(\mathbf{y}) d\mathbf{y} \le \int_{W(\mathbf{y}) \le s_2} Q(\mathbf{y}) p(\mathbf{y}) d\mathbf{y}.$$

Now, replacing $\{\mathbf{y} : \widetilde{Q}(\mathbf{y}) \le s_1 - Q_0\}$ with the original $\{\mathbf{y} : Q(\mathbf{y}) \le s_1\}$ proves inequality (1.27) when Q may be negative but bounded from below.

Finally, consider the general case when Q may be negative and is not bounded from below. Define a sequence of functions $\widetilde{Q}_k(\mathbf{y}) = \max(k, Q(\mathbf{y}))$ where $k = -1, -2, \dots$. Due to continuity of p and Q, and the existence of $E(Q)$ there exist $\varepsilon_k \to 0$ such that

$$\int_{Q(\mathbf{y}) \le s_1} p(\mathbf{y}) d\mathbf{y} = \int_{\widetilde{Q}_k(\mathbf{y}) \le s_1 + \varepsilon_k} p(\mathbf{y}) d\mathbf{y} = \int_{W(\mathbf{y}) \le s_2} p(\mathbf{y}) d\mathbf{y},$$

which implies

$$\int_{\widetilde{Q}_k(\mathbf{y}) \le s_1 + \varepsilon_k} \widetilde{Q}_k(\mathbf{y}) p(\mathbf{y}) d\mathbf{y} \le \int_{W(\mathbf{y}) \le s_2} \widetilde{Q}_k(\mathbf{y}) p(\mathbf{y}) d\mathbf{y}.$$

because $\widetilde{Q}_k$ is bounded from below. Also, due to the continuity, there exist δ_{k1} and $\delta_{k2} \to 0$ such that

$$\int_{\widetilde{Q}_k(\mathbf{y}) \le s_1 + \varepsilon_k} \widetilde{Q}_k(\mathbf{y}) p(\mathbf{y}) d\mathbf{y} = \int_{Q_k(\mathbf{y}) \le s_1} Q(\mathbf{y}) p(\mathbf{y}) d\mathbf{y} + \delta_{k1},$$

$$\int_{W(\mathbf{y}) \le s_2} \widetilde{Q}_k(\mathbf{y}) p(\mathbf{y}) d\mathbf{y} = \int_{W(\mathbf{y}) \le s_2} Q(\mathbf{y}) p(\mathbf{y}) d\mathbf{y} + \delta_{k2}.$$

Finally, letting $k \to -\infty$, we arrive at the inequality (1.27). $\square$

The result of this lemma can be expressed as the solution of a linear optimization problem in function space under a linear constraint: the minimum of the integral over the Lebesgue measurable set M,

$$\min_M \int_M Q(\mathbf{y}) p(\mathbf{y}) d\mathbf{y},$$

under the restriction $\int_M p(\mathbf{y}) d\mathbf{y} = r$, where $r > 0$, is attained at $M = \{\mathbf{y} : Q(\mathbf{y}) \le s\}$, where s is such that $\int_{Q(\mathbf{y}) \le s} p(\mathbf{y}) d\mathbf{y} = r$. It can be easily proved for a one-dimensional case using the Lagrange multiplier technique. Reduce the search for the optimal $M \subset R^1$ to an interval (a, b). Then the optimization problem reduces to

$$\min_{a<b} \int_a^b Q(y)p(y)dy \text{ under constraint } \int_a^b p(y)dy = r.$$

Introduce the Lagrange function

$$\mathcal{L}(a, b; \lambda) = \int_a^b Q(y)p(y)dy - \lambda \left(\int_a^b p(y)dy - r \right),$$

where λ is a Lagrange multiplier. Differentiate with respect to a and b

$$\frac{\partial \mathcal{L}}{\partial a} = -Q(a)p(a) + \lambda p(a) = 0, \quad \frac{\partial \mathcal{L}}{\partial a} = Q(b)p(b) - \lambda p(b) = 0$$

to find the optimal solution $Q(a) = Q(b)$, that is, the value of Q at the both sides of the optimal interval is the same.

Next we deduce an important familiar result that among the sets of the same area, the level set has maximum volume under the density.

Corollary 1.6 *Let $p(\mathbf{y})$ be a positive continuous pdf and $B \subset R^n$ be a compact set. Then*

$$\int_{\{p(\mathbf{y}) \geq \kappa\}} d\mathbf{y} = \int_B d\mathbf{y}, \quad \kappa > 0 \tag{1.31}$$

implies

$$\int_{\{p(\mathbf{y}) \geq \kappa\}} p(\mathbf{y})d\mathbf{y} \geq \int_B p(\mathbf{y})d\mathbf{y}. \tag{1.32}$$

Proof. We reformulate the corollary claim in the language of Lemma 1.5. However, instead of the original equations, we use the equivalent equations (1.28) and (1.29). First, define a uniform pdf p_U on the union of the sets $\{p(\mathbf{y}) \geq \kappa\}$ and B as

$$p_U(\mathbf{y}) = \begin{cases} A^{-1} \text{ for all } \mathbf{y} \text{ such that } p(\mathbf{y}) \geq \kappa \text{ or } \mathbf{y} \in B \\ 0 \text{ elsewhere} \end{cases},$$

where $A = \int_{\{p(\mathbf{u}) \geq \kappa\} \cup B} d\mathbf{u}$ is the volume of the union. Since $\kappa > 0$ and B is compact the union set is compact and has a finite volume, $A < \infty$. Second, let Q act as p. It is easy to see that $E(Q) < \infty$ because

$$E(Q) = A^{-1} \int_{\{p(\mathbf{u}) \geq \kappa\} \cup B} p(\mathbf{y})d\mathbf{u} < \infty.$$

Third, define set B via a function $W(\mathbf{y})$ as a set of points $\{W(\mathbf{y}) \geq s_2\}$. Then equation (1.31) can be rewritten in the form (1.28) where p is replaced with p_U and $\{Q(\mathbf{y}) \geq s_1\}$ is replaced with $\{p(\mathbf{y}) \geq \kappa\}$ as

$$\int_{\{p(\mathbf{y}) \geq \kappa\}} p_U(\mathbf{y}) d\mathbf{y} = \int_{\{W(\mathbf{y}) \geq s_2\}} p_U(\mathbf{y}) d\mathbf{y}.$$

Now we rewrite equation (1.29) via Q and p_U as

$$\int_{\{p(\mathbf{y}) \geq \kappa\}} p(\mathbf{y}) p_U(\mathbf{y}) d\mathbf{y} \geq \int_B p(\mathbf{y}) p_U(\mathbf{y}) d\mathbf{y}.$$

But on the sets $\{p(\mathbf{y}) \geq \kappa\}$ and B, the pdf p_U takes the value A^{-1} and therefore the previous inequality turns into

$$\frac{1}{A} \int_{\{p(\mathbf{y}) \geq \kappa\}} p(\mathbf{y}) d\mathbf{y} \geq \frac{1}{A} \int_B p(\mathbf{y})) d\mathbf{y},$$

which is equivalent to the desired (1.32). □

We prove this result for the finite discrete distribution of random variable Y. Although the original distribution may be multidimensional without loss of generality we may assume that the discrete random variable is defined on the real line at the set of points/outcomes $\{y_1, y_2, ..., y_M\}$ with probabilities $\{p_1, p_2, ..., p_M\}$, where $p_m > 0$ and $\sum_{m=1}^{M} p_m = 1$. The set B is defined as a subset of distinct outcomes, that is, $\{y_m^*, m = 1, 2, ..., M^*\}$ with probabilities $\{p_m^*, m = 1, 2, ..., M^*\}$, where $M^* \leq M$ and each y_m^* is one of $\{y_1, y_2, ..., y_M\}$. The volume of B is the number of points in this subset, M^*. The inequality $p(\mathbf{y}) \geq \kappa$ defines the second subset of outcomes $\{y_m^{**}, m = 1, ..., M^*\}$ with probabilities $\{p_m^{**}, m = 1, 2, ..., M^*\}$ such $p_m^{**} \geq \kappa$. We want to show that

$$\sum_{m=1}^{M^*} p_m^{**} \geq \sum_{m=1}^{M^*} p_m^*. \tag{1.33}$$

Indeed, if there exists y_m^* with $\Pr(Y = y_m^*) = p_m < \kappa$, we replace y_m^* with y_m^{**} for which $p_m^{**} \geq \kappa$. This substitution keeps the volume, that is, M^* the same but increases $\sum_{m=1}^{M^*} p_m^*$. We can continue this substitution for any other y_m^* for which $p_m < \kappa$. This proves (1.33).

The following result on the integral inequality is dual to the previous lemma. It has a familiar interpretation: among all sets with the same volume under the density the density level set $\{p(\mathbf{y}) \geq c\}$ has the minimal area. In statistics interpretation, it proves that the density level test introduced in Section 2.3 has the minimum volume acceptance region.

Lemma 1.7 *Let $p(\mathbf{y})$ be a positive continuous pdf and W be a continuous function of $\mathbf{y} \in R^n$ bounded from below, $W(\mathbf{y}) > w_0$. Then*

$$\int_{p(\mathbf{y}) \geq c_1} p(\mathbf{y}) d\mathbf{y} = \int_{W(\mathbf{y}) \geq c_2} p(\mathbf{y}) d\mathbf{y} \qquad (1.34)$$

implies

$$\int_{p(\mathbf{y}) \geq c_1} d\mathbf{y} \leq \int_{W(\mathbf{y}) \geq c_2} d\mathbf{y} \qquad (1.35)$$

where c_1 and $c_2 > w_0$ are any constants.

Proof. Note that without loss of generality, we can assume that function W is positive by replacing it in the domain of integration with $W - w_0$ and c_2 with $c_2 - w_0$. Now define $\widetilde{Q}_k(\mathbf{y}) = 1/p(\mathbf{y})$ if $p(\mathbf{y}) \geq 1/k$ and 0 elsewhere where $k = 1, 2, \ldots$. Then there exists a sequence $\varepsilon_k \to 0$ such that

$$\int_{p(\mathbf{y}) \geq c_1} p(\mathbf{y}) d\mathbf{y} = \int_{\widetilde{Q}_k(\mathbf{y}) \leq 1/c_1 + \varepsilon_k} p(\mathbf{y}) d\mathbf{y}$$

for all k starting from some k_0. Indeed, denote the left-hand side as $A = \text{const}$, and consider function $f_k(\varepsilon) = \int_{\widetilde{Q}_k(\mathbf{y}) \leq 1/c_1 + \varepsilon} p(\mathbf{y}) d\mathbf{y}$ for nonnegative ε. This function is increasing and continuous. Obviously, $f_k(\infty) = 1 > A$. Moreover, starting from some $k_0 \leq k$ we have $f_k(0) < A$. This means for some $f_k(\varepsilon_k) = A$, and therefore such a sequence of ε_k exists. Hence condition (1.34) can be equivalently written as

$$\int_{\widetilde{Q}_k(\mathbf{y}) \leq 1/c_1 + \varepsilon_k} p(\mathbf{y}) d\mathbf{y} = \int_{1/W(\mathbf{y}) \leq 1/c_2} p(\mathbf{y}) d\mathbf{y}. \qquad (1.36)$$

According to Lemma 1.5, it implies that for all $k \geq k_0$ we have

$$\int_{\widetilde{Q}_k(\mathbf{y}) \leq 1/c_1 + \varepsilon_k} \widetilde{Q}_k(\mathbf{y}) p(\mathbf{y}) d\mathbf{y} \leq \int_{1/W(\mathbf{y}) \leq 1/c_2} \widetilde{Q}_k(\mathbf{y}) p(\mathbf{y}) d\mathbf{y}. \qquad (1.37)$$

By the choice of ε_k, (1.36) is equivalent to (1.34). Let $k \to \infty$ in (1.37). We aim to prove that the left- and the right-hand side of (1.37) converge to the respective sides of (1.35). Indeed, when $k \to \infty$ for the right-hand side of (1.37) we have

$$\int_{W(\mathbf{y}) \geq c_2} \widetilde{Q}_k(\mathbf{y}) p(\mathbf{y}) d\mathbf{y} = \int_{\{W(\mathbf{y}) \geq c_2\} \cap \{p(\mathbf{y}) \geq 1/k\}} d\mathbf{y} \to \int_{W(\mathbf{y}) \geq c_2} d\mathbf{y}.$$

Note that in the above limit, we can safely assume that the set $\{W(\mathbf{y}) \geq c_2\}$ is bounded and the integral at right is finite because if $\{W(\mathbf{y}) \geq c_2\}$ is unbounded the inequality (1.35) holds since the right-hand side is infinity. Similarly, for the left-hand-side of (1.37) letting $k \to \infty$ we have

$$\int_{\widetilde{Q}_k(\mathbf{y}) \leq 1/c_1 + \varepsilon_k} \widetilde{Q}_k(\mathbf{y}) p(\mathbf{y}) d\mathbf{y} = \int_{p(\mathbf{y}) \geq \max(c_1/(1 + c_1 \varepsilon_k, 1/k)} d\mathbf{y} \to \int_{p(\mathbf{y}) \geq c_1} d\mathbf{y}.$$

This proves that (1.34) implies (1.35). $\qquad\qquad\qquad\qquad\qquad\qquad$ □

Following, we illustrate the previous lemmas with a one-dimensional case ($n = 1$) (Casella and Berger 1990; Demidenko 2020). In particular, we emphasize the duality of the two optimization problems.

Proposition 1.8 *Suppose that pdf p is differentiable and unimodal on $(-\infty, \infty)$. (a) Among all intervals with the same area under the pdf the short interval has the same pdf values at the ends of the interval. (b) Among all intervals of fixed length, the area under the pdf attains a maximum when the pdf values at the ends of the interval are the same.*

Proof. (a) If the area is $0 < A < 1$, then

$$\int_{-\infty}^{T} p(x) dx - \int_{-\infty}^{t} p(x) dx = A. \qquad (1.38)$$

Find t and T by solving the optimization problem $\min(T - t)$ under constraint (1.38). Introduce the Lagrange function

$$\mathcal{L}(t, T; \lambda) = T - t - \lambda \left(\int_{-\infty}^{T} p(x) dx - \int_{-\infty}^{t} p(x) dx - A \right),$$

where λ is the Lagrange multiplier. The necessary conditions for the minimum are

$$\frac{\partial \mathcal{L}}{\partial t} = -1 + \lambda p(t) = 0, \quad \frac{\partial \mathcal{L}}{\partial T} = 1 - \lambda p(T) = 0.$$

These equations imply $p(t) = p(T)$. Since f is a unimodal equation, $f(t) = 1/\lambda$ has two solutions, $t < T$.
    (b) Now we want to maximize

$$\int_{-\infty}^{T} p(x) dx - \int_{-\infty}^{t} p(x) dx$$

under constraint $T - t = a$. The Lagrange function takes the form

$$\mathcal{L}(t, T; \lambda) = \int_{-\infty}^{T} p(x) dx - \int_{-\infty}^{t} p(x) dx T - \lambda(T - t - a).$$

Differentiation yields

$$\frac{\partial \mathcal{L}}{\partial t} = -p(t) + \lambda = 0, \quad \frac{\partial \mathcal{L}}{\partial T} = p(T) - \lambda = 0,$$

so again $p(t) = p(T)$. □

Remarks:

1. The two optimization problems, (a) and (b), are dual to each other and lead to equivalent Lagrange functions.

2. Figure 1.7 illustrates Proposition 1.8: the area under the pdf for both intervals (bold and dotted horizontal segments) is the same, but the interval with the same values of the pdf at the ends of the interval is the shortest.

3. The condition in equation (1.38) can be expressed as

$$F(T) - F(t) = A,$$

where F is the cdf. Consequently, we may rephrase Proposition 1.8 as follows: (a) among all intervals with the same coverage probability the short interval (bold) has the same pdf values at the ends of the interval; (b) among all intervals of fixed length the coverage probability attains maximum when the pdf values at the ends of the interval are the same.

We will be using the results from this section to prove the optimal properties of the tests to be developed in the following part of the book. For example, we will use Lemma 1.7 to prove that the density level test has the minimum volume of the acceptance region.

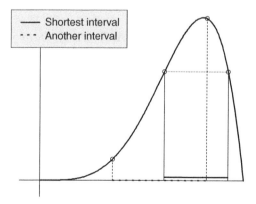

Figure 1.7: *Illustration of Proposition 1.8. The shortest interval (bold segment) has the same area under the pdf but a smaller length than another/dotted interval on the x-axis.*

References

[1] Anderson, R.L. and Bancroft, T.A. (1952). *Statistical Theory in Research*. New York: McGraw-Hill.

[2] Bickel, P.J. and Doksum, K.A. (2001). *Mathematical Statistics: Basic Ideas and Selected Topics*, vol. 1. Prentice Hall.

[3] Casella, G. and Berger, R.L. (1990). *Statistical Inference*. Belmont, CA: Duxbury Press.

[4] Chen, S. (2003). A class of parameter functions for which the unbiased estimator does not exist. *Lecture Notes-Monograph Series, Institute of Mathematical Statistics* 43: 159–164.

[5] Cox, D.R. and Hinkley, D.V. (1996). Theoretical Statistics. Boca Raton: CRC Press.

[6] Demidenko, E. (2007). Sample size determination for logistic regression revisited. *Statistics in Medicine* 26: 3385–3397.

[7] Demidenko, E. (2020). *Advanced Statistics with Applications in R*. Hoboken, NJ: Wiley.

[8] Deming, W.E. (1964). *Statistical Adjustment of Data*. New York: Dover.

[9] El-Bassiouni, M.Y. (1994). Short confidence intervals for variance components. *Communications in Statistics – Theory and Methods* 23: 1915–1933.

[10] Fieller, E.C. (1932). The distribution of the index in a normal bivariate population. *Biometrika* 24: 428–440.

[11] Ghosh, M. (1996). On the nonexistence of nonnegative unbiased estimators of variance components. *Sankhyā: The Indian Journal of Statistics, Series B* 58: 360–362.

[12] Hogg, R.V., McKean, J.W., and Craig A.T. (2006). *Introduction to Mathematical Statistics*. Upper Saddle River, NJ: Pearson Education.

[13] Lehmann, E.L. and Casella, G. (1998). *Theory of Point Estimation*, 2e. New York: Springer.

[14] Lehmann, E.L. and Romano, J.P. (2005). *Testing Statistical Hypothesis*, 3e. New York: Springer.

[15] Neyman, J. and Pearson, E.S. (1933). On the problem of the most efficient tests of statistical hypothesis. *Philosophical Transactions of the Royal Society A* 231: 694–706.

[16] Neyman, J. and Pearson, E.S. (1936). Contributions to the theory of testing statistical hypotheses I. *Statistical Research Memoirs* 1: 1–37.

[17] Ortega, J.M. and Rheinboldt, W.C. (2000). *Iterative Solution of Nonlinear Equations in Several Variables*. Philadelphia: SIAM.

[18] Rao, C.R. (1973). *Linear Statistical Inference and Its Applications*, 2e. New York: Wiley.

[19] Schervish, M. (2012). *Theory of Statistics*. Springer: New York.

[20] Silverman, B.W. (1986). *Density Estimation*. London: Chapman and Hall.

[21] Shao, J. (2003). *Mathematical Statistics*. New York: Springer.

[22] Tate, R.F. and Klett, G.W. (1959). Optimal confidence intervals for the variance of a normal distribution. *Journal of the American Statistical Association* 54: 674–682.

[23] Tukey, J.W. (1977). *Exploratory Data Analysis*. Addison-Wesley.

[24] Zacks, S. (1971). *The Theory of Statistical Inference*. New York: Wiley.

Chapter 2

Maximum concentration statistics

The goal of this chapter is to present general results on maximum concentration (MC) statistics for the one-dimensional parameter. The multidimensional case is discussed in Chapter 6. First, we formulate the necessary assumptions. Then we introduce the minimum length (short) two-sided confidence interval (CI) and the implied MC estimator as the limit point when the confidence level goes to zero. In the spirit of MC-statistics, we derive the density level (DL) test as the test with the short acceptance interval. Finally, we discuss the efficiency and its connection with sufficient statistics. We use the term maximum *concentration statistics* because the MC estimator is where the concentration probability of the coverage of the true parameter reaches its maximum.

2.1 Assumptions

Our M-statistics relies on the existence of statistic S having cdf $F(s; \theta)$ known up to unknown one-dimensional parameter θ. Here we assume that S is a continuous random variable. Moreover, we assume that the pdf $f(s; \theta)$ is positive for all $s \in (-\infty, \infty)$ and unimodal for every θ. In the following sections, we apply M-statistics to discrete distributions.

Three cases will be distinguished: when the parameter belongs to the entire real line, is positive, or belongs to an open finite interval such as $(0, 1)$. Classic statistics treats these cases under one theoretical umbrella. This creates a problem with an unbiased estimation of a positive parameter because then estimates should take negative values to produce the unbiasedness of the parameter close to zero. For example, this problem emerges in the unbiased

M-statistics: Optimal Statistical Inference for a Small Sample, First Edition. Eugene Demidenko.
© 2023 John Wiley & Sons, Inc. Published 2023 by John Wiley & Sons, Inc.

estimation of the variance components (meta-analysis) model and squared correlation coefficient (see Sections 5.9 and 5.11). In this section, we assume that the parameter belongs to the entire line, $\theta \in (-\infty, \infty)$. The case when $\theta > 0$ or belongs to $(0, 1)$ is discussed in Section 2.5

The following properties of F are required (F' and F'' denote differentiation with respect to the parameter). For every $s \in (-\infty, \infty)$, we demand:

1. $F(s; \theta)$ is a continuously differentiable function with respect to θ up to the second order.

2. $F'(s; \theta) < 0$.

3. $F''(s; \theta) = 0$ has a unique solution for θ, the inflection point on the cdf curve as a function of θ.

4. $\lim_{\theta \to -\infty} F(s; \theta) = 1$ and $\lim_{\theta \to \infty} F(s; \theta) = 0$.

These properties hold for many but not all distributions. For example, property 4 does not hold for the coefficient of determination – see Section 5.12.

We illustrate properties 1–4 with location parameter θ when statistic S has cdf $F(s; \theta) = H((s - \theta)/\nu)$, where H is a known cdf with the unimodal differentiable positive density h and ν is a known scale parameter. Since $F'(s; \theta) = -\nu^{-1} h((s - \theta)/\nu) < 0$, the second condition holds. The third condition holds as well with the unique inflection point $s - \nu M$, where M is the mode of the H-distribution. It is easy to see that the fourth condition holds as well.

When the parameter is positive, $-\infty$ is replaced with 0 – see the following example as a clarification.

Example 2.1 *Assumptions 1–4 hold for normal variance.*

For normal variance, we choose $S = \sum_{i=1}^{n} (Y_i - \overline{Y})^2$ with F as the cdf of the chi-square distribution with $n - 1$ df, $F(s; \sigma^2) = C_{n-1}(s/\sigma^2)$, where $0 < \sigma^2 < \infty$. Obviously, the first assumption holds. Since

$$\frac{d}{d\sigma^2} C_{n-1}\left(\frac{s}{\sigma^2}\right) = -\frac{1}{\sigma^4} c_{n-1}\left(\frac{s}{\sigma^2}\right) < 0$$

the second assumption holds as well. For the third assumption we have

$$\frac{d^2}{d\sigma^4} C_{n-1}\left(\frac{s}{\sigma^2}\right)$$
$$= \frac{2}{\sigma^6} c_{n-1}\left(\frac{s}{\sigma^2}\right) - \frac{1}{\sigma^4} c_{n-1}\left(\frac{s}{\sigma^2}\right) \frac{d}{d\sigma^2} \ln c_{n-1}\left(\frac{s}{\sigma^2}\right)$$

$$= \frac{2}{\sigma^6} c_{n-1}\left(\frac{s}{\sigma^2}\right) - \frac{1}{\sigma^4} c_{n-1}\left(\frac{s}{\sigma^2}\right)\left(-\frac{n-3}{2\sigma^2} + \frac{s}{2\sigma^4}\right)$$

$$= \frac{1}{2\sigma^6} c_{n-1}\left(\frac{s}{\sigma^2}\right)\left(n + 1 - \frac{s}{\sigma^2}\right).$$

This means the second derivative turns into zero at $s/(n+1)$ – the inflection point on the curve $C_{n-1}(s/\sigma^2)$ as a function of σ^2. This solution corresponds to the MC estimator of σ^2 as discussed earlier in Section 1.3.3. The fourth assumption holds because $C_{n-1}(s/0) = C_{n-1}(\infty) = 1$.

Example 2.2 *Assumptions 3–4 do not hold for a uniform distribution.*

Consider the estimation of the upper limit of the uniform distribution $\theta > 0$ having n iid observations $X_i \sim \mathcal{R}(0, \theta)$. An obvious statistic is $S = \max X_i$ with the cdf $F(s; \theta) = (s/\theta)^n$. Since $F''(s; \theta) = n(n+1)s^n\theta^{-(n+2)}$, the third assumption does not hold – there is no inflection point of the curve $F(s; \theta)$. The fourth assumption does not hold either – when $\theta \to 0$, we have $F(s; \theta) \to \infty$.   □

The following result is fundamental for the construction of exact CIs.

Theorem 2.3 *Let λ be a given number from the interval $(0, 1)$ and $L(s)$ and $U(s)$ be strictly increasing continuous functions such that*

$$F(s; L(s)) - F(s; U(s)) = \lambda \quad \forall s, \tag{2.1}$$

where F satisfies Assumptions 1–4. Then

$$\Pr(L(S) \leq \theta \leq U(S)) = \lambda,$$

where S is the observed statistic having cdf F with parameter θ.

Proof. Due to the conditions of the theorem the inverse functions L^{-1} and U^{-1} exist. Then

$$\Pr(L(S) \leq \theta \leq U(S)) = \Pr(L(S) \leq \theta) - \Pr(U(S) < \theta)$$
$$= \Pr(S \leq L^{-1}(\theta)) - \Pr(S < U^{-1}(\theta)) = F(L^{-1}(\theta); \theta) - F(U^{-1}(\theta); \theta).$$

Letting $s = L^{-1}(\theta)$ and $s = U^{-1}(\theta)$ rewrite $F(L^{-1}(\theta); \theta) = F(s; L(s))$ and $F(U^{-1}(\theta); \theta) = F(s; U(s))$. Now using these substitutions due to the law of repeated expectation and condition (2.1), we finally obtain

$$\Pr(L(S) \leq \theta \leq U(S)) = E_S[(F(S; L(S)) - F(S; U(S)))|S = s]$$
$$= E_S(\lambda) = \lambda,$$

where E_S is the expectation over S and $|$ indicates the conditional distribution.   □

Functions L and U can be viewed as one-to-one mappings of the observation to parameter space. Note that his theorem remains true when the parameter space is a finite, such as $0 < \theta < 1$, or a semi-infinite, such as $\theta > 0$.

If F satisfies the previous four assumptions, and L and U satisfy the conditions of Theorem 2.3 and S is the observed statistic the CI $(L(S), U(S))$ covers the true parameter θ with the exact probability λ. In traditional settings, the equal-tail CIs are used by finding functions L and U by solving $F(s; L(s)) = (1 + \lambda)/2$ and $F(s; U(s)) = (1 - \lambda)/2$. It is easy to prove that functions L and U are increasing. Indeed, by differentiating both sides of the first equation with respect to s, we obtain

$$\frac{d}{ds} F(s; L(s)) = f(s; L(s)) + F'(s; L(s)) L'(s) = 0,$$

where $f > 0$ is the pdf. Since $F' < 0$, we infer that $L'(s) > 0$. Similarly, we prove that U is an increasing function. Conditions for increasing functions L and U in the unequal-tail approaches in MC and MO-statistics frameworks are provided in Theorem 2.6 and Theorem 3.6, respectively.

Importantly, L and U are not defined uniquely from condition (2.1). In M-statistics, we improve the equal-tail CI by a special choice of L and U using additional optimality criteria for the CI to be short or unbiased.

2.2   Short confidence interval and MC estimator

The goal of this section is to derive the minimum-length (short) CI and implied MC estimator. The basis for our theory is the CI derived from the inverse (with respect to parameter) cdf. Casella and Berger (1990, p. 416) attribute this method to Mood et al. (1974). This method can be viewed as a special case of the pivotal CI (Shao 2003) with the pivotal quantity $F(S; \theta) \sim \mathcal{R}(0, 1)$. See sections 5.9 and 6.8 for more detail. We will refer to this method as the *inverse cdf* that reflects the essence of the method (Demidenko 2020).

The development of the short CI has a long history, starting from the seminal work by Neyman and Pearson (1936). Short CIs were developed for some particular statistical problems in the framework of the general pivotal quantity by Guenther (1972), Dahiya and Guttman (1982), Ferentinos (1988), Juola (1993), Ferentinos and Karakostas (2006), and a few others. Our goal is to construct the general theory for the short CI using the cdf of statistic S known up to the unknown parameter θ.

Let $0 < \lambda < 1$ be the desired confidence level. Following the method of inverse cdf, given the observed statistic S, we define random variables $\theta_L < \theta_U$ that cover the true parameter θ with probability λ. In terms of the cdf, the coverage statement can be equivalently expressed as $F(S; \theta_L) - F(S; \theta_U) = \lambda$. In addition, we want to make our CI as short as possible by minimizing $\theta_U - \theta_L$. To find the confidence limits, introduce the Lagrange function similarly to the normal variance example,

$$\mathcal{L}(\theta_L, \theta_U, \nu) = \theta_U - \theta_L - \nu[F(S; \theta_L) - F(S; \theta_U) - \lambda].$$

The necessary condition for the optimum yields the second equation $F'(S; \theta_L) = F'(S; \theta_U)$. Finally, the exact two-sided CI for θ with minimum length is defined as (θ_L, θ_U), where $\theta_L = \theta_L(S)$ and $\theta_U = \theta_U(S)$ are the solutions of the system of equations

$$F(S; \theta_L) - F(S; \theta_U) - \lambda = 0, \quad F'(S; \theta_L) - F'(S; \theta_U) = 0. \qquad (2.2)$$

We shall call this CI the *short CI*. The existence and uniqueness of the solution for fixed S and λ are given in Theorem 2.6.

We need to make an important comment on the difference between the minimum expected length CI, the previously known from the literature (Pratt 1961), and the short CI defined by (2.2). While the former has the minimal *expected* length the latter has minimum length for each value of statistic S: that is, its length is *uniformly* smaller than any other CI as a function of S with the fixed coverage probability.

We make an important distinction between the unequal-tail CI derived from (2.2) and the equal-tail CI often discussed in the literature where θ_{L0} and θ_{U0} are obtained from equations

$$F(S; \theta_{L0}) = (1 + \lambda)/2, \ F(S; \theta_{U0}) = (1 - \lambda)/2.$$

Although both intervals have the same coverage probability the length of the CI with confidence limits obeying (2.2) is smaller. The system (2.2) is solved by Newton's algorithm with the adjustment vector for (θ_L, θ_U) given by

$$\begin{bmatrix} -F'(S; \theta_L) & F'(S; \theta_U) \\ F''(S; \theta_L) & -F''(S; \theta_U) \end{bmatrix}^{-1} \begin{bmatrix} F(S; \theta_U) - F(S; \theta_L) - \lambda \\ F'(S; \theta_L) - F'(S; \theta_U) \end{bmatrix},$$

starting from the equal-tail confidence limits, θ_{L0} and θ_{U0}. Typically, it requires three or four iterations to converge.

Looking ahead, note that the short CI is an optimal CI on the *additive* scale because the size of the interval is measured as $\theta_U - \theta_L$, in contrast to the case when the parameter is positive with the size measured on the *relative* scale, say, as $\ln(\theta_U/\theta_L)$ – see Section 2.5 for details.

Now we appeal to Lemma 1.7 to derive the short CI from a different angle. Define function $\kappa = \kappa(\theta)$ such that

$$\int_{\{f(s;\theta) \geq \kappa\}} f(s; \theta) ds = \lambda, \qquad (2.3)$$

where f is the density. Then, given the observed value S, the CI (θ_L, θ_U) is the level set $\{\theta : f(S; \theta) \geq \kappa(\theta)\}$. The fact that this CI has the minimal length among all intervals that cover the true θ with probability λ follows from Lemma 1.7. Indeed, any other CI can be expressed as the level set of a function W dependent on s and θ. Since it is required to cover the true θ with probability λ we have the condition

$$\int_{\{f(s;\theta)\geq\kappa\}} f(s;\theta)ds = \int_{\{W(s;\theta)\geq\kappa_2\}} f(s;\theta)ds$$

But from the lemma, we infer that the length of the density level CI is not greater than the length of the CI derived from function W for any θ because

$$\int_{\{f(s;\theta)\geq\kappa\}} ds \leq \int_{\{W(s;\theta)\geq\kappa_2\}} ds.$$

One can avoid the two-step procedure by noting that the lower and upper limits of the CI can be derived by solving the equation

$$\int_{\{f(s;\theta)\geq f(S;\theta)\}} f(s;\theta)ds = \lambda. \tag{2.4}$$

Indeed, since f is unimodal the domain of integration can be expressed as an interval (S, s_*), where $f(s_*; \theta) = f(S; \theta)$, where without loss of generality we assume that the equation $f(s; \theta) = f(S; \theta)$, besides an obvious $s = S$, has another solution $s_* > S$. Then (2.4) can be written as $F(s_*; \theta) - F(S; \theta) = \lambda$. Since F is invertible, we may find θ_L such that $F(S; \theta_L) = F(s_*; \theta)$. By letting $\theta = \theta_U$, we finally conclude that the solution of (2.4) is equivalent to (3.4). The method of finding the boundary of the CI by solving (2.4) will be generalized to the multidimensional parameter in Section 6.1.

Example 2.4 *Short CI for normal variance and standard deviation.*

We apply the CI (2.2) to the normal variance from the previous section where $F(S; \sigma^2) = C_{n-1}(S/\sigma^2)$. Since in our case $F'(S; \sigma^2) = -S/\sigma^4 c_{n-1}(S/\sigma^2)$, equations (2.2) can be equivalently written as

$$C_{n-1}(S/\sigma_L^2) - C_{n-1}(S/\sigma_U^2) - \lambda = 0,$$
$$(S/\sigma_L^2)^2 c_{n-1}(S/\sigma_L^2) - (S/\sigma_U^2)^2 c_{n-1}(S/\sigma_U^2) = 0. \tag{2.5}$$

Letting $q_U = S/\sigma_L^2$ and $q_L = S/\sigma_U^2$, we arrive at equations (1.2) and (1.5) solved by Newton's algorithm realized in the R function `var.ql` with `adj=-1`. Thus, the short CI (2.2) yields the minimum expected length CI for the normal variance.

Now we apply (2.2) to find the CI for standard deviation σ with $F(S; \sigma) = C_{n-1}(S/\sigma^2)$ that leads to the system of equations

$$C_{n-1}(q_U) - C_{n-1}(q_L) - \lambda = 0, \quad q_L^{3/2} c_{n-1}(q_L) - q_U^{3/2} c_{n-1}(q_U) = 0.$$

Again, this system can be effectively solved by Newton's algorithm. Once q_L and q_U are found, the CI for σ takes the form $(\sqrt{S/q_U}, \sqrt{S/q_L})$. Note that the limits of the CI for the standard deviation (SD) are not equal to the square roots of the limits of the variance. Comprehensive statistical inference for the SD is offered in Section 5.1.

Definition 2.5 *The maximum concentration (MC) estimator, $\widehat{\theta}_{MC} = \widehat{\theta}_{MC}(S)$, is the limit point of the short CI (2.2) when $\lambda \to 0$.*

The following theorem proves (a) the existence and uniqueness of system (2.2), and (b) that the MC estimator is the inflection point of F as a function of θ, that is, the point of the steepest slope.

Theorem 2.6 *If the cdf $F(s; \theta)$ satisfies properties 1–4, then (a) the system of equations (2.2) has a unique solution $\theta_L < \theta_U$ for every observed S and $\lambda \in (0,1)$. (b) When $\lambda \to 0$ interval (θ_L, θ_U) converges to $\widehat{\theta}_{MC} = \widehat{\theta}_{MC}(S)$, the solution of the equation*

$$F''(S; \theta) = 0. \tag{2.6}$$

That is, the MC estimate is the inflection point on the curve $F(S; \theta)$, or equivalently where $F(S; \theta)$ has the steepest (negative) slope. (c) For every confidence level λ, interval (θ_L, θ_U) contains $\widehat{\theta}_{MC}$.

Proof. Since S is fixed, we omit it for brevity. We start with an obvious observation that the first and second derivatives of F vanish when θ approaches $\pm\infty$ due to property 4. Also, it is easy to see that $F'(\theta)$ is a decreasing function to the left and an increasing function to the right of $\widehat{\theta}_{MC}$ due to property 3. (a) Express θ_L through θ_U from the first equation of (2.2) as $\theta_L = F^{-1}(F(\theta_U) + \lambda)$ under condition $F(\theta_U) + \lambda < 1$, where F^{-1} is the θ-inverse. Note that the inverse function, F^{-1}, is twice continuously differentiable. Substitute this expression for θ_L into the second equation of (2.2), and define a differentiable function

$$G(\theta_U) = F'(F^{-1}(F(\theta_U) + \lambda)) - F'(\theta_U), \quad \theta_U > F^{-1}(1 - \lambda).$$

Then the uniqueness and existence of the solution (2.2) is equivalent to the uniqueness and existence of the solution $G(\theta_U) = 0$. The solution of $G(\theta_U) = 0$ exists because

$$\lim_{\theta_U \to \infty} G(\theta_U) = F'(F^{-1}(\lambda)) < 0,$$

$$\lim_{\theta_U \to F^{-1}(1-\lambda)} G(\theta_U) = -F'(F^{-1}(1 - \lambda)) > 0.$$

The uniqueness of $\theta_L < \theta_U$ follows from property 3 and representation $G(\theta_U) = F'(\theta_L) - F'(\theta_U)$. (b) From the first equation of (2.2), it follows that

$\lim_{\lambda \to 0} \theta_L = \lim_{\lambda \to 0} \theta_U = \theta$. From the second equation it follows

$$0 = \lim_{\lambda \to 0} \frac{F'(\theta_L) - F'(\theta_U)}{\theta_L - \theta_U} = F''(\theta),$$

which means the MC estimate obeys (2.6). (c) The interval (θ_L, θ_U) contains $\widehat{\theta}_{MC}$ because for $\theta < \widehat{\theta}_{MC}$, we have $F''(\theta) < 0$, and for $\theta > \widehat{\theta}_{MC}$, we have $F''(\theta) > 0$.   □

The fact that the MC estimator is the point of the steepest slope on $F(S; \theta)$ explains why we call it the *maximum concentration* estimator, because letting $\lambda \to 0$, we have

$$\frac{\Pr(\theta_L < \theta < \theta_U)}{\theta_U - \theta_L} = \frac{F(S; \theta_L) - F(S; \theta_U)}{\theta_U - \theta_L} \to F'(S; \widehat{\theta}_{MC}). \qquad (2.7)$$

One may interpret the left-hand side as the *concentration* probability, that is, the probability of the coverage of the true parameter per the length of the interval. Thus we say that the MC estimator maximizes the infinitesimal concentration probability.

Figure 2.1 illustrates the MC estimator of the normal variance with $n = 10$. It depicts $F(S; \sigma^2) = C_{n-1}(S/\sigma^2)$ as a function of σ^2 (bold curve) with the observed (random) value $S = 14.79$. The dotted line is $F'(S; \sigma^2)$ with the values shown at the y-axis at right with the observed $S = 14.79$. Traditional unbiased variance estimate $\widehat{\sigma}^2 = 14.79/9 = 1.64$. The MC estimate, $\widehat{\sigma}^2_{MC} = 1.34$, is where $F'(S; \sigma^2)$ is minimal, the point of the deepest descent, or the inflection point on the $F(S; \sigma^2)$ curve, shown as the circle.

Example 2.7 *The MC estimator of the normal variance.*

The MC estimator for σ^2 was derived earlier (1.13) by letting $\lambda \to 0$. Now we use the condition (2.6). Since $F(s; \sigma^2) = C_{n-1}(S/\sigma^2)$, elementary calculus yields the equation

$$F''(S; \sigma^2) = \sigma^{-6} c_{n-1}(S/\sigma^2)(2 + \sigma^2(n-3)/(2\sigma^2) - S/(2\sigma^2)) = 0.$$

This equation is equivalent to $2 + (n-3)/2 - S/(2\sigma^2) = 0$. Simple algebra produces the solution given by (1.13).

Example 2.8 *Location parameter.*

The MC estimator of the location parameter θ with a known scale parameter ν defined by the cdf $F(s; \theta) = H((s - \theta)/\nu)$ is $\widehat{\theta}_{MC} = S - \nu M$, where M is the mode. Note that for this problem, the MC estimator coincides with the maximum likelihood estimator if S is a sufficient statistic. Indeed, then the likelihood

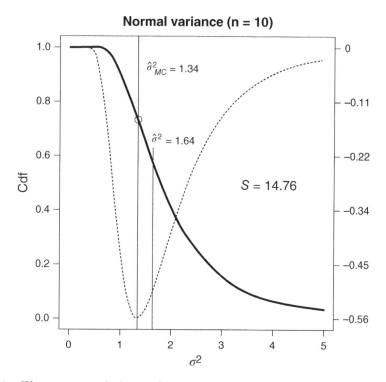

Figure 2.1: *Illustration of the MC estimator of the normal variance, $n = 10$. The cdf $F(S; \sigma^2) = C_{n-1}(S/\sigma^2)$ is plotted against σ^2 with the circle indicating the MC estimate of the normal variance as the point of the steepest descent. The dotted line is $F'(S; \theta)$.*

function is $L(\theta) = \nu^{-1} h((S - \theta)/\nu)$, which takes a maximum at the mode, and therefore $\widehat{\theta}_{ML} = S - \nu M = \widehat{\theta}_{MC}$.

Example 2.9 *Exponential distribution.*

Let Y_i be independent exponentially distributed observations with unknown positive scale parameter θ. The traditional unbiased estimator of θ is $\widehat{\theta} = \overline{Y}$. Statistic $S = \sum_{i=1}^{n} Y_i$ follows a chi-square distribution with $2n$ df, that is, $F(s; \theta) = C_{2n}(2s/\theta)$. Equating the second derivative of the cdf with respect to θ to zero, after some algebra, we obtain

$$\widehat{\theta}_{MC} = \frac{1}{n+1} \sum_{i=1}^{n} Y_i.$$

Figure 2.2 illustrates the computation of $\widehat{\theta}_{MC}$ as the point of the deepest slope on the cdf curve as a function of θ. The bold line depicts $F(S; \theta)$, and the dotted line depicts $F'(S; \theta)$ with $S = 2.11$ and the minimum at $\widehat{\theta}_{MC} = 0.53$.

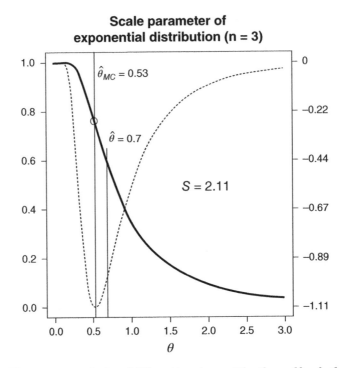

Figure 2.2: *Illustration of the MC estimator with the cdf of the exponential distribution $F(S; \theta) = G_n(S/\theta)$ for computation of the MC estimate of the scale parameter in the exponential distribution – see Example 2.9.*

The unbiased estimate of the scale parameter is $\widehat{\theta} = 0.7$. The superiority of the MC estimator for small n in terms of proximity to the true θ is illustrated in Figure 2.3. Define the proximity to the true θ in terms of the concentration probabilities $\delta^{-1} \Pr(|\widehat{\theta} - \theta| < \delta/2)$ and $\delta^{-1} \Pr(|\widehat{\theta}_{MC} - \theta| < \delta/2)$ as functions of δ. Two concentration probabilities for $n = 3$ and $n = 6$ with the true scale parameter $\theta = 3$ are shown. The MC estimate has a higher concentration around the true scale parameter θ than the classic unbiased estimate on the entire range of δ. The circles at $\delta = 0$ represent the density of $\widehat{\theta}$ and $\widehat{\theta}_{MC}$ evaluated at the true θ as the limit of the concentration probability when $\delta \to 0$. Interestingly, for $n = 3$, the concentration probabilities are decreasing functions, but for $n = 6$ are increasing functions of δ.

Note that the exponential distribution parametrized by the rate λ with the cdf $1 - e^{-\lambda s}$ does not satisfy Assumption #2 because $d(1 - e^{-\lambda s})/d\lambda = se^{-\lambda s} > 0$.

Example 2.10 *Order statistic for the location parameter.*

Consider estimation of the location parameter with the cdf given by $H(s - \theta)$, where $h = H'$ is the unimodal differentiable density (the scale parameter, $\nu = 1$).

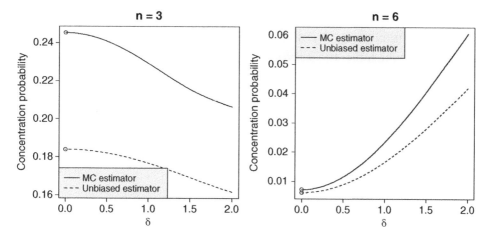

Figure 2.3: *Concentration probabilities for the MC and classic unbiased estimator of the scale parameter θ in an exponential distribution, Example 2.9. The MC estimate has a higher concentration around the true scale parameter than the classic unbiased estimates.*

If $\{X_i, i = 1, ..., n\}$ is a random sample from this distribution choose the order statistic $S = X_{(k)}$ with the cdf evaluated at S given by

$$F(S; \theta) = \sum_{j=k}^{n} \binom{n}{j} H^j(S - \theta)(1 - H(S - \theta))^{n-j},$$

where $1 \le k \le n$. Take the derivative to obtain the pdf:

$$F' = -h \left[\sum_{j=k}^{n-1} \binom{n}{j} (j - nH)H^{j-1}(1 - H)^{n-j-1} + nH^{n-1} \right],$$

where $x = S - \theta$ is omitted for brevity. If x_m is the minimizer of $-F'(x; \theta)$, then the MC estimator is given by $\widehat{\theta}_{MC} = X_{(k)} - x_m$, with $x_{(m)}$ as an adjustment. In a special case when $n = 3$ and $k = 2$, we have $F'(x; \theta) = -6hH(1 - H)$. If the distribution is symmetric, the minimum of the derivative is at $x_m = 0$. Therefore, the adjustment is zero and $\widehat{\theta}_{MC} = X_{(2)}$. Of course, for an asymmetric distribution, the adjustment is not equal to zero.

Example 2.11 *Cauchy distribution with an unknown location parameter*.

The MC estimator and the short CI can be derived using the order statistic as in Example 2.10, where $H(x) = \pi^{-1} \arctan x + 0.5$. To illustrate, suppose that $n = 3$ and only the maximum observation is available, $X_{(3)}$. Then the steepest slope reduces to the minimization of function $-F' = (\pi^{-1} \arctan x + 0.5)^2/(1 + x^2)$ with the minimizer $x_m = 0.493$. Therefore, $\widehat{\theta}_{MC} = X_{(3)} - 0.493$. Now we use the average $\overline{X}$, which has the same cdf F, and therefore $\widehat{\theta}_{MC} = \overline{X}$ with the symmetric λ confidence level CI for θ given by $\overline{X} \pm \tan(\pi\lambda/2)$. Interestingly, the minimum variance estimator does not exist for the Cauchy distribution (Freue 2007). Note that the well-known invariant Pitman estimator does not apply in these two examples because the mean and variance do not exist (Zacks 1971).

2.3　Density level test

The goal of this section is to introduce the *density level (DL)* test as the test with a short acceptance interval. The multidimensional version of the test is discussed in Section 6. According to the DL test, the null hypothesis $H_0 : \theta = \theta_0$ is accepted if $f(S; \theta_0) \geq \kappa_0$, where κ_0 is such that $f(q_L; \theta_0) = f(q_U; \theta_0) = \kappa_0$ under condition $F(q_U; \theta_0) - F(q_L; \theta_0) = \lambda$. Alternatively, since f is unimodal at θ_0, the acceptance interval is $q_L \leq S \leq q_U$. Formally, to derive q_L and q_U, one replaces S with q_U and q_L as the first argument, and q_U and q_L with θ_0 as the second argument. Accordingly, the derivative of the cdf with respect to the second argument must be replaced with the derivative with respect to the first argument which produces the density that leads to the system of equations

$$F(q_U; \theta_0) - F(q_L; \theta_0) = \lambda, \quad f(q_L; \theta_0) - f(q_U; \theta_0) = 0. \tag{2.8}$$

To find q_L and q_U, Newton's algorithm applies, with the adjustment vector at each iteration given by

$$\begin{bmatrix} -f(q_L; \theta_0) & f(q_U; \theta_0) \\ f'_s(q_L; \theta_0) & -f'_s(q_U; \theta_0) \end{bmatrix}^{-1} \begin{bmatrix} F(q_U; \theta_0) - F(q_L; \theta_0) - \lambda \\ f(q_L; \theta_0) - f(q_U; \theta_0) \end{bmatrix},$$

where the subscript s indicates differentiation of the pdf $f(s; \theta)$ with respect to s. A good start is the equal-tail values as the $(1 - \lambda)/2$ and $(1 + \lambda)/2$ quantiles.

Theorem 2.12 *If, in addition to properties 1–4 from Section 2.1,*

$$\frac{d^2}{ds^2} \ln f(s; \theta_0) < 0 \quad \forall s, \tag{2.9}$$

system (2.8) has a unique solution for every $0 < \lambda < 1$.

Proof. Let $\lambda = 1 - \alpha$, and omit the second argument θ_0 for brevity. Express q_U through $s = q_L$ from the first equation of (2.8) as $q_U = Q(\lambda + F(s))$, where Q is the quantile function. Define the left-hand side of the second equation of (2.8) as a function of s

$$H(s) \stackrel{\text{def}}{=} f(s) - f(Q(F(s) + \lambda)), \quad s < Q(\alpha).$$

Prove that equation $H(s) = 0$ has at least one solution and the solution is unique on the interval $s < Q(\alpha)$. Indeed,

$$\lim_{s \to -\infty} H(s) = f(-\infty) - f(Q(\lambda)) < 0,$$

$$\lim_{s \to Q(\alpha)} H(s) = f(Q(\alpha)) - f(\infty) > 0$$

proves that the solution exists. Now prove that the solution of $H(s) = 0$ is unique by taking the derivative of H using the chain rule and noting that $Q' = 1/f$:

$$\frac{dH}{ds} = f'(s) - \frac{f'(Q(F(s) + \lambda))}{f(Q(F(s) + \lambda))} f(s)$$

$$= f(s) \left(\frac{d}{ds} \ln f(s) - \frac{d}{ds} \ln f(s_*) \right),$$

where $s_* = Q(F(s) + \lambda) > s$. But due to property 6 function $d \ln f(s)/ds$ is decreasing and therefore the expression in parentheses is negative. This proves that system (3.1) has a unique solution. $\qquad\square$

The following property is elementary to prove and yet underscores the optimality of the DL test with the short acceptance interval.

Theorem 2.13 *Among all exact tests based on the test statistic S, the DL test has the shortest acceptance interval.*

Proof. We aim to prove that among all pairs $q_L < q_U$ such that $F(q_U; \theta_0) - F(q_L; \theta_0) = \lambda$ the pair derived from (2.8) has minimum $q_U - q_L$. Similarly to the short CI introduce the Lagrange function

$$\mathcal{L}(q_L, q_U, \nu) = q_U - q_L - \nu[F(q_U; \theta_0) - F(q_L; \theta_0) - \lambda].$$

Differentiation with respect to q_L and q_U yields system (2.8). $\qquad\square$

The name "density level" can be justified in connection with Lemma 1.7 and specifically Proposition 1.8 as follows. Find κ_0 such that

$$\int_{\{f(s;\theta_0) \ge \kappa_0\}} f(s; \theta_0) ds = \lambda. \tag{2.10}$$

Then the null hypothesis is accepted if S belongs to the level set $\{s : f(s; \theta_0) \ge \kappa_0\}$: that is, $f(S; \theta_0) \ge \kappa_0$. This formulation of the DL test will be used for the multidimensional parameter in Chapter 5.

It is easy to see that (2.8) and (2.10) are just different formulations of the same test. Indeed, since f is unimodal the level set $\{f(s; \theta_0) \ge \kappa_0\}$ turns into an interval (q_L, q_U) with the equal density values at the ends, $f(q_L; \theta_0) = f(q_U; \theta_0)$.

Clearly (2.10) can be rewritten as $F(q_U; \theta_0) - F(q_L; \theta_0) = \lambda$, which proves that the two formulations are equivalent.

As follows from Lemma 1.7, the DL test is optimal in the sense that it has the tightest acceptance region, that is, the shortest acceptance interval among all exact tests with the test statistic S. This optimal property is an implication of the following well-known result: among all intervals with the same area under the pdf, the shortest interval has equal values of the pdf at the ends.

The integral formulation (2.10) is useful for generalization to the multidimensional parameter to be discussed in Section 6.1.

The DL test has an obvious similarity to the classic likelihood ratio test, however, there are differences. The first difference is that the classic test involves the joint pdf of all observations while we use the pdf of statistic S. The second difference is that the likelihood ratio test typically relies on the asymptotic chi-square distribution and implied approximate size, but the DL test has the exact size.

The power function of the DL test is given by

$$P(\theta; \theta_0) = 1 + F(q_L; \theta) - F(q_U; \theta), \tag{2.11}$$

where q_L and q_U dependent on θ_0 are found from (2.8). Indeed, since the distribution of $F(S; \theta)$ under the alternative θ is uniform on $(0, 1)$, the power function, as the probability of rejecting the null, is

$$\begin{aligned}
P(\theta; \theta_0) &= 1 - \Pr(q_L < S < q_U) \\
&= 1 - \Pr(F(q_L; \theta) < F(S; \theta) < F(q_U; \theta)) \\
&= 1 + F(q_L; \theta) - F(q_U; \theta).
\end{aligned}$$

The DL test is consistent: that is, when θ approaches $-\infty$ or ∞, the power approaches 1, as follows from property 4. This test is generally biased because the derivative of the power function with respect to θ evaluated at θ_0 may not de zero, as illustrated in the following example (the unbiased test is developed in Section 6.1).

Example 2.14 *Density level and dual tests for normal variance.*

For normal variance, $F(s; \sigma^2) = C_{n-1}(s/\sigma^2)$. The solution of system (2.8) for the DL test boils down to finding quantiles q_L and q_U that satisfy the following equations:

$$C_{n-1}(q_U) - C_{n-1}(q_L) - \lambda = 0, \quad (n-3)(\ln q_U - \ln q_L) - (q_U - q_L) = 0. \tag{2.12}$$

This system is solved numerically in R by the function var.ql using Newton's algorithm with adj=3. The power function computed by (2.11) is depicted in

Figure 1.5. The derivative of P with respect to σ^2 evaluated at σ_0^2 is negative – the test is biased. Note that the test dual to the short CI is different from the DL test. As follows from Section 1.3.2, the test dual to the short CI for the variance from Example 2.4 uses q_L and q_U, found from function `var.ql` with `adj=-1`, but the DL test uses q_L and q_U with `adj=3`. The difference between the tests is seen in Figure 1.5.

2.4 Efficiency and the sufficient statistic

This section answers an obvious question – what statistic is best for the MC inference, and how does one measure the efficiency of the MC estimator? In the classic minimum variance unbiased estimation approach the Rao-Blackwell theorem says that the best unbiased estimator is a function of a sufficient statistic. As we shall learn from this section, a similar answer holds for the MC inference.

An obvious feature of merit of the MC estimator is the slope/derivative of the parent cdf because the steeper the slope the higher the concentration probability of the coverage of the true parameter – see the justification around equation (2.7). In M-statistics, we compare the efficiency of estimators on the cdf scale (this holds for MO estimators as well to be discussed later) – two estimators are compared under the condition that the cdf values are the same, therefore the following definition.

Definition 2.15 *We say that for a given parameter value θ the MC estimator based on statistic S is no less MC efficient than another MC estimator based on statistic R with cdfs F_S and F_R, respectively, if $F_S(s_1; \theta) = F_R(s_2; \theta)$ implies $F_S'(s_1; \theta) \leq F_R'(s_2; \theta)$. We say that statistic S is* uniformly *no less MC efficient if it is MC efficient for every parameter value θ.*

In other words, among all statistics with the same value of the cdf, an efficient statistic produces an MC estimator with a steeper cdf slope. The inequality $F_S'(s_1; \theta) \leq F_R'(s_2; \theta)$ may be tested on the quantile scale for $p \in (0, 1)$, where s_1 and s_2 are pth quantiles of cdfs $F_S(s_1; \theta)$ and $F_R(s_2; \theta)$. This representation is especially convenient when plotting the slope of the cdfs against p for comparison – see Example 2.17.

Here we prove that a sufficient statistic yields the most efficient MC estimator – Lemma 1.5 is instrumental for the proof. Looking back, the expected efficiency of the sufficient statistic supports Definition 2.15.

Theorem 2.16 *The MC optimality of sufficient statistic. Let $S = S(\mathbf{y})$ be a sufficient statistic: that is, the pdf can be expressed in the form $f(\mathbf{y}; \theta) = e^{g(S(\mathbf{y}); \theta) + p(\theta) + h(\mathbf{y})}$, where $g_\theta'(s; \theta) = \partial g / \partial \theta$ is a strictly increasing function of s. Then the MC estimator based on statistic S is no less MC-efficient than the MC estimator based on another statistic $R = R(\mathbf{y})$.*

Proof. Express the cdfs of statistics, F_S and F_R in an integral form. Then the condition on equal cdf values at s_1 and s_2 can be written as

$$F_S(s_1; \theta) = \int_{S(\mathbf{y}) \leq s_1} f(\mathbf{y}; \theta) d\mathbf{y} = \int_{R(\mathbf{y}) \leq s_2} f(\mathbf{y}; \theta) d\mathbf{y} = F_R(s_2; \theta). \qquad (2.13)$$

Since the derivative of F_S can be expressed as

$$F_S'(s_1; \theta) = \int_{S(\mathbf{y}) \leq s_1} (g_\theta'(S(\mathbf{y}); \theta) + p'(\theta)) f(\mathbf{y}; \theta) d\mathbf{y}$$

$$= \int_{S(\mathbf{y}) \leq s_1} g_\theta'(S(\mathbf{y}); \theta) f(y; \theta) d\mathbf{y} + p'(\theta) \int_{S(\mathbf{y}) \leq s_1} f(\mathbf{y}; \theta) d\mathbf{y}$$

and similarly the derivative of F_R, thanks to condition (2.13), we suffice to prove that

$$F_S'(s_1; \theta) = \int_{S(\mathbf{y}) \leq s_1} g_\theta'(S(\mathbf{y}); \theta) f(\mathbf{y}; \theta) d\mathbf{y}$$

$$\leq \int_{R(\mathbf{y}) \leq s_2} g_\theta'(S(\mathbf{y}); \theta) f(\mathbf{y}; \theta) d\mathbf{y} = F_R'(s_2; \theta). \qquad (2.14)$$

Since $g_\theta'(s; \boldsymbol{\theta})$ is a strictly increasing function with respect to s, the inequalities that define the integration domain in R^n can be equivalently expressed in terms of g_θ' as

$$\{S(\mathbf{y}) \leq s_1\} = \{g_\theta'(S(\mathbf{y}); \theta) \leq g_\theta'(s_1; \theta)\},$$

$$\{R(\mathbf{y}) \leq s_2\} = \{g_\theta'(R(\mathbf{y}); \theta) \leq g_\theta'(s_2; \theta)\}.$$

Define $\quad Q(\mathbf{y}) = g_\theta'(S(\mathbf{y}); \theta), \quad W(\mathbf{y}) = g_\theta'(R(\mathbf{y}); \theta), \quad$ and $\quad s_1^* = g_\theta'(s_1; \theta),$ $s_2^* = g_\theta'(s_2; \theta)$ to match the notation of Lemma 1.5. Then (2.13) yields the familiar condition

$$\int_{Q(\mathbf{y}) \leq s_1^*} p(\mathbf{y}) d\mathbf{y} = \int_{W(\mathbf{y}) \leq s_2^*} p(\mathbf{y}) d\mathbf{y},$$

where $p(\mathbf{y}) = f(\mathbf{y}; \theta)$. But as follows from Lemma 1.5, it implies

$$\int_{Q(\mathbf{y}) \leq s_1^*} Q(\mathbf{y}) p(\mathbf{y}) d\mathbf{y} \leq \int_{W(\mathbf{y}) \leq s_2^*} Q(\mathbf{y}) p(\mathbf{y}) d\mathbf{y},$$

which in turn produces the required inequality (2.14). $\qquad \square$

We remark that usually, the pdf due to the Neyman factorization theorem, is written in the form $G(S(\mathbf{y}); \theta, \boldsymbol{\tau}) H(\mathbf{y})$. On a technical note, we use an equivalent exponential representation $e^{g(S(\mathbf{y}); \boldsymbol{\theta}) + p(\theta) + h(\mathbf{y})}$. The term $p(\theta)$ relaxes the condition on the strictly increasing function g_θ', because $p(\theta)$ may be any – see Example 2.17 for an illustration.

Example 2.17 *Normal variance with known mean.*

The purpose of this example is to show that for normal variance when $\mu = \mu_0$ is known, the sufficient statistic produces an efficient MC estimator compared to a non-sufficient statistic in view of Definition 2.15. First, we show that the conditions of Theorem 2.16 hold. Indeed, it is easy to see that $g(S; \sigma^2) = -(2\sigma^2)^{-1}S$, $p(\sigma^2) = -(n/2)(\ln \sigma^2 + \ln \sigma^2)$ and $H(\mathbf{y}) = (2\pi)^{-n/2}$, where $S = \sum_{i=1}^n (Y_i - \mu_0)^2$ is the sufficient statistic with cdf $F_S(s; \sigma^2) = C_n(s/\sigma^2)$. It is elementary to see that $g'_{\sigma^2}(S; \sigma^2) = S\sigma^{-4}/2$ is an increasing function of S, and therefore the condition of Theorem 2.16 holds. To find the S-based MC estimator we solve the equation $C''_n(S/\sigma^2) = 0$. Omitting some elementary algebra we obtain $\widehat{\sigma}^2_{MC} = S/(n-1)$. Now we turn our attention to a competitor, $R = \sum_{i=1}^n (Y_i - \overline{Y})^2$, which is not a sufficient statistic with cdf $F_R(s; \sigma^2) = C_{n-1}(s/\sigma^2)$. This statistic gives birth to another MC estimator $\tilde{\sigma}^2_{MC} = R/(n+1)$. We aim to show that the former estimate produces a steeper cdf when the two estimates have the same cdf value: that is, $C_n(s_1/\sigma^2) = C_{n-1}(s_2/\sigma^2)$ implies $C'_n(s_1/\sigma^2) = -s_1/\sigma^4 c_n(s_1/\sigma^2) < -s_2/\sigma^4 c_{n-1}(s_2/\sigma^2) = C'_{n-1}(s_2/\sigma^2)$. Letting $\tau_1 = s_1/\sigma^2$ and $\tau_2 = s_2/\sigma^2$, we claim that $C_n(\tau_1) = C_{n-1}(\tau_2)$ yields $\tau_1 c_n(\tau_1) > \tau_2 c_{n-1}(\tau_2)$. Figure 2.4 provides a visual confirmation that the cdf curve for statistic S has a steeper slope than statistic

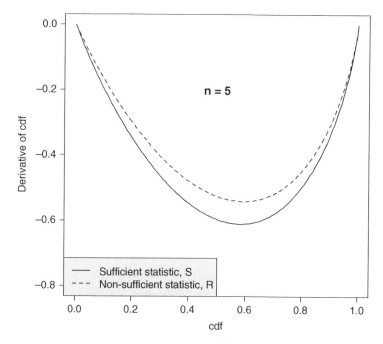

Figure 2.4: *The plot of the derivative $F'(s; \sigma^2)$ versus the cdf $F(s; \sigma^2)$ for a sufficient statistic S and a non-sufficient statistic R in Example 2.17. If the cdf values of S and R are the same the slope of the cdf for the sufficient statistic is steeper.*

R if they have the same cdf values – the solid line is below the dotted line on the entire range of the cdf values.

Example 2.18 *Sufficient statistic for an exponential family.*

The pdf of a single observation Y takes the form $f(y; \theta) = e^{r(\theta)K(y) - p(\theta) + h(y)}$, where $r'(\theta) > 0$. With n iid observations, the pdf of the sample is given by $f(\mathbf{y}; \theta) = e^{r(\theta)S - np(\theta) + h(\mathbf{y})}$, where $h(\mathbf{y}) = \sum_{i=1}^{n} h(y_i)$ and $S = \sum_{i=1}^{n} K(Y_i)$ is the sufficient statistic. In the notation of Theorem 2.16, we have $g'(S; \theta) = r'(\theta)S$ being a strictly increasing function of S, and therefore we conclude that the sufficient statistic S gives birth to the MC-efficient estimator of θ.

2.5 Parameter is positive or belongs to a finite interval

As mentioned previously, the classic statistics does not take into account the parameter space domain. Consequently, it creates a problem with zero or negative estimates when the parameter is positive. This section offers a solution for staying away from zero values by replacing the traditional additive scale with the log or double-log scale.

2.5.1 Parameter is positive

If the parameter is positive the additive scale is not appropriate – the length of the CI must be measured on the relative scale. Indeed, when the true θ_0 is positive and small the length of the interval, $\theta_U - \theta_L$, cannot be fairly compared with the case when θ_0 is large. Several relative measures can be considered. The variance example, to be considered later, suggests the log scale by minimizing $\ln(\theta_U / \theta_L) = \ln \theta_U - \ln \theta_L$. An advantage of the log scale is that the interval and the respective estimator stay away from zero – a common problem with the traditional approach.

Following the argumentation from the previous section, the Lagrange multiplier technique yields the short CI on the log scale as the solution of (2.2) with the second equation replaced with

$$\theta_L F'(S; \theta_L) - \theta_U F'(S; \theta_U) = 0. \tag{2.15}$$

To guarantee the existence and uniqueness of the CI and the respective MC estimator on the log scale (MCL) as the limit point of CI when $\lambda \to 0$, conditions on the cdf listed in Section 2.1 must be modified. The first two requirements remain the same but the last two are replaced with the following. When the parameter is positive, $\theta > 0$, for every s as the value of statistic S, instead of Assumptions 3 and 4 from Section 2.1, we demand:

3. The equation $F'(s; \theta) + \theta F''(s; \theta) = 0$ has a unique solution for θ as the minimizer of $\theta F'(s; \theta)$.

4. $\lim_{\theta \to \infty} F(s; \theta) = \lim_{\theta \to 0} \theta F'(s; \theta) = \lim_{\theta \to \infty} \theta F'(s; \theta) = 0$ and $\lim_{\theta \to 0} F(s; \theta) = 1$.

The examples that follow clarify the assumptions and computation of the CI.

Example 2.19 *CI for the normal variance on the log scale.*

We prove that the optimal CI for the normal variance on the log scale defined by (2.15) turns into an unbiased CI (1.10). Indeed, since the cdf of S is $F(s; \sigma^2) = C_{n-1}(s/\sigma^2)$, we have $\sigma^2 F'(S; \sigma^2) = -S/\sigma^2 c_{n-1}(S/\sigma^2)$, and therefore condition (2.15) turns into (1.10) with $q_L = S/\sigma_U^2$ and $q_U = S/\sigma_L^2$.

Definition 2.20 *The maximum concentration estimator on the log scale (MCL), $\widehat{\theta}_{MCL} = \widehat{\theta}_{MCL}(S)$, is the limit point of the short CI (2.15) when $\lambda \to 0$.*

Similarly to Theorem 2.6, under the previous assumptions, the MCL estimate, $\widehat{\theta}_{MCL}$, exists, is unique and positive, and is found by solving the equation

$$F'(S; \theta) + \theta F''(S; \theta) = 0.$$

Geometrically, $\widehat{\theta}_{MCL}$ is where the curve $\theta F'(S; \theta)$ has the steepest negative slope.

The MCL estimator is the point of the highest concentration of the coverage probability on the relative scale and penalizes the MC estimator to be close to zero. Indeed, it is easy to see that $\widehat{\theta}_{MC} < \widehat{\theta}_{MCL}$, because $F'(s; \theta) + \theta F''(s; \theta)$ evaluated at $\theta = \widehat{\theta}_{MC}$ is negative when θ is close to zero and $F''(s; 0) < 0$.

The following examples demonstrate that some traditional estimators can be viewed as MCL estimators.

Example 2.21 *Normal variance (continued).*

Elementary differentiation gives

$$F''(s; \sigma^2) = s \left[2/\sigma^2 + (n-3)/(2\sigma^2) - s/(2\sigma^4) \right] \sigma^{-4} c_{n-1}(s/\sigma^2).$$

Hence $F'(s; \theta) + \theta F''(s; \theta) = 0$ leads to a unique solution $\sigma^2 = s/(n-1)$, and therefore $\sigma^2_{MCL} = \widehat{\sigma}^2$, the traditional unbiased estimator (1.14).

Example 2.22 *Exponential distribution (continued).*

In Example 2.9 of Section 2.2, it was shown that if the scale parameter was not restricted to be positive, $\widehat{\theta}_{MC} = S/(n+1)$. Now we use the fact that θ is positive and measure the length of the CI on the log scale. Then if $\lambda \to 0$, the CI shrinks to the solution of $F'(S;\theta) + \theta F''(S;\theta) = 0$. Some elementary calculus yields $\widehat{\theta}_{MCL} = \overline{Y}$, the traditional unbiased estimator.

Example 2.23 *Normal standard deviation.*

The same statistic $S = \sum (Y_i - \overline{Y})^2$ is used to derive the MCL of σ with cdf $F(s;\sigma) = C_{n-1}(s/\sigma^2)$. Elementary differentiation yields

$$F'(s;\sigma) = -2s\sigma^{-3}c_{n-1}(s/\sigma^2),$$

$$F''(s;\sigma) = s\left(\frac{6}{\sigma^4} + \frac{4s}{\sigma^6}\left(\sigma^2\frac{n-3}{2s} - \frac{1}{2}\right)\right)c_{n-1}(s/\sigma^2).$$

The MCL estimator is the solution of $F'(S;\sigma) + \sigma F''(S;\sigma) = 0$, and after some algebra, one derives a commonly used estimator,

$$\widehat{\sigma} = \widehat{\sigma}_{MCL} = \sqrt{S/(n-1)}.$$

Remarkably, everybody uses $\widehat{\sigma}$ although an unbiased estimator exists, $\tilde{\sigma} = \sqrt{S/\nu_n}$, where $\nu_n = 2\Gamma^2(n/2)/\Gamma^2((n-1)/2)$; see Johnson et al. (1994). Our MCL theory supports the traditional estimator as the limit point of the short CI on the log scale when the confidence level approaches zero.

2.5.2   Parameter belongs to a finite interval

Sometimes a parameter such as probability, belongs to an open finite interval. To simplify the discussion, we shall assume that $\theta \in (0,1)$. Then, following the previous argumentation, we need to address both possibilities: when the true parameter is close to the left or right end. We propose to measure the length of the CI on the *double-log scale*:

$$(\ln \theta_U - \ln \theta_L) + (\ln(1 - \theta_L) - \ln(1 - \theta_U)).$$

This criterion penalizes the confidence limits for being close to 0 or 1. The Lagrange multiplier technique yields the solution for the short CI on the log scale:

$$F(S;\theta_L) - F(S;\theta_U) - \lambda = 0,$$

$$\theta_L(1 - \theta_L)F'(S;\theta_L) - \theta_U(1 - \theta_U)F'(S;\theta_U) = 0. \qquad (2.16)$$

To obtain the MC estimator on the double-log scale (MCL2), let $\lambda \to 0$, $\theta_L \to \theta$, and $\theta_U \to \theta$ that implies

$$0 = \lim_{\theta_L \to \theta,\ \theta_U \to \theta} \frac{\theta_L(1-\theta_L)F'(S;\theta_L) - \theta_U(1-\theta_U)F'(S;\theta_U)}{\theta_L - \theta_U}$$
$$= (1-2\theta)F'(S;\theta) + \theta(1-\theta)F''(S;\theta).$$

Thus the MCL2 estimator, $\widehat{\theta}_{MCL2}$, is derived by solving the equation

$$(1-2\theta)F'(S;\theta) + \theta(1-\theta)F''(S;\theta) = 0, \tag{2.17}$$

or, equivalently, the minimizer of the function $\theta(1-\theta)F'(S;\theta)$.

The maximum concentration estimator on the log scale resolves the paradox of the binomial probability and Poisson rate with zero outcomes – see Section 5.7 and Section 5.8. More applications of M-statistics on the log scale are found in Section 5.9 and Section 5.11.

References

[1] Casella, G. and Berger, R.L. (1990). *Statistical Inference*. Belmont, CA: Duxbury Press.

[2] Dahiya, R. and Guttman, I. (1982). Shortest confidence intervals and prediction intervals for the log-normal. *Canadian Journal of Statistics* 10: 277–291.

[3] Demidenko, E. (2020). *Advanced Statistics with Applications in R*. Hoboken, NJ: Wiley.

[4] Ferentinos, K. (1988). On shortest confidence intervals and their relation with uniformly minimum variance unbiased estimators. *Statistical Papers* 29: 59–75.

[5] Ferentinos, K.K. and Karakostas, K.X. (2006). More on shortest and equal tails confidence intervals. *Communications in Statistics – Theory and Methods* 35: 821–829.

[6] Freue, G.V.C. (2007). The Pitman estimator of the Cauchy location parameter. *Journal of Statistical Planning and Inference* 137: 1900–1913.

[7] Guenther, W.C. (1972). On the use of the incomplete gamma table to obtain unbiased tests and unbiased confidence intervals for the variance of the normal distribution. *The American Statistician* 26: 31–34.

[8] Johnson, N.L, Kotz, S., and Balakrishnan, N. (1994). *Continuous Univariate Distributions*, vol. 1, 2e. New York: Wiley.

[9] Juola, R.C. (1993). More on shortest confidence intervals. *The American Statistician* 47: 117–119.

[10] Mood, A.M. (1974). *Introduction to the Theory of Statistics*. New York: McGraw-Hill.

[11] Neyman, J. and Pearson, E.S. (1936). Contributions to the theory of testing statistical hypotheses I. *Statistical Research Memoirs* 1: 1–37.

[12] Pratt, J.W. (1961). Length of confidence intervals. *Journal of the American Statistical Association* 56: 549–567.

[13] Shao, J. (2003). *Mathematical Statistics*. New York: Springer.

[14] Zacks, S. (1971). *The Theory of Statistical Inference*. New York: Wiley.

Chapter 3

Mode statistics

This chapter presents an alternative to maximum concentration statistics where the short CI is replaced with the unbiased CI and the MC estimator is replaced with the MO estimator. We start with an introduction of the unbiased test and unbiased CI. Then the respective mode (MO) estimator is derived as the limit point of the unbiased CI when the confidence level approaches zero. Finally, in the framework of MO-statistics, we demonstrate the optimal property of the sufficient statistic through the concept of cumulative information as a generalization of classic Fisher information.

3.1   Unbiased test

The notion of the unbiased test was introduced by Neyman and Pearson many years ago. Unbiased tests are intuitively appealing because the probability of rejecting the null is minimal at the null value (Casella and Berger 1990, Shao 2003, Lehmann and Romano 2005). Consequently, the probability of rejecting the hypothesis at the parameter value different from the null is greater than at the null value. Typically, equal-tail quantiles are used to guarantee the test size α. We argue that with a special choice of quantiles, we can make a test unbiased. This section suggests the quantiles that turn the derivative of the test evaluated at the null hypothesis to zero which yields a locally unbiased test. Sufficient conditions are offered that guarantee that the quantiles exist and that the test is globally unbiased.

The unbiased double-sided test for normal variance discussed in our motivating example of Section 1.3 is easy to generalize. To test the null hypothesis $H_0 : \theta = \theta_0$ versus the alternative $H_0 : \theta \neq \theta_0$, we start by finding

M-statistics: Optimal Statistical Inference for a Small Sample, First Edition. Eugene Demidenko.
© 2023 John Wiley & Sons, Inc. Published 2023 by John Wiley & Sons, Inc.

quantiles $q_L < q_U$ such that

$$F(q_U; \theta_0) - F(q_L; \theta_0) = 1 - \alpha, \quad F'(q_L; \theta_0) - F'(q_U; \theta_0) = 0, \qquad (3.1)$$

where differentiation is understood with respect to the second argument (parameter) evaluated at θ_0. The first equation guarantees that the size of the test is α, and the second equation guarantees that the test is unbiased; see the details in Theorem 3.1. The hypothesis is rejected if the observed S falls outside of the interval (q_L, q_U). In terms of the cdf, the null hypothesis is accepted if

$$F(q_L; \theta_0) \leq F(S; \theta_0) \leq F(q_U; \theta_0).$$

The formula for the power function of the unbiased test is the same (2.11) but now with q_L and q_U obtained from (3.1). The local unbiasedness of the test follows directly from the differentiation of the power function

$$\left. \frac{d}{d\theta} P(\theta; \theta_0) \right|_{\theta=\theta_0} = F'(q_L; \theta_0) - F'(q_U; \theta_0) = 0$$

due to the second equation (3.1). Global unbiasedness, under an additional condition, is proven next.

Theorem 3.1 *If in addition to properties 1–4 from Section 2.1,*

$$\frac{\partial^2 \ln f(s; \theta_0)}{\partial s \partial \theta} > 0 \quad \forall s, \qquad (3.2)$$

the following holds: (a) the system (3.1) has a unique solution for every $0 < \lambda < 1$; (b) the power function (2.11) is increasing at right and decreasing at left of θ_0 and approaches 1 when $\theta \to \pm\infty$, and therefore the test is globally unbiased and consistent; (c) the test is unique among all unbiased exact tests with the test statistic as a function of S.

Proof. (a) Let $\lambda = 1 - \alpha$, and express q_U through $s = q_L$ from the first equation (θ_0 is omitted for brevity hereafter). After plugging it into the second equation of (3.1) define function

$$H(s) = F'(s) - F'(Q(F(s) + \lambda)), \quad s < Q(\alpha),$$

where Q is the quantile function, that is, $F(Q(s)) = s$. The existence and uniqueness of solution of the system (3.1) are equivalent to the existence and uniqueness of solution of the equation $H(s) = 0$ on the interval $s < Q(\alpha)$. To prove the existence we find the limits of H at the extremes:

$$\lim_{s \to -\infty} H(s) = 0 - F'(Q(\lambda)) > 0, \quad \lim_{s \to Q(\alpha)} H(s) = F'(Q(\alpha)) - 0 < 0.$$

This implies that system (3.1) has at least one solution. Now we prove that $H(s)$ is a strictly decreasing function. Indeed,

$$\frac{dH}{ds} = f'(s) - \frac{f'(Q(F(s) + \lambda))}{f(Q(F(s) + \lambda))} f(s) = f(s) \left(\frac{d}{d\theta} \ln f(s) - \frac{d}{d\theta} \ln f(s_*) \right),$$

where $s_* = Q(F(s) + \lambda) > s$. But from the condition (3.2), it follows that function $d \ln f(s; \theta_0)/d\theta$ is an increasing function of s, and therefore the expression in the parentheses is negative. This proves that system (3.1) has a unique solution.

(b) Since

$$\frac{dP(\theta; \theta_0)}{d\theta} = F'(q_L; \theta_0) - F'(q_U; \theta_0)$$

and solution $F'(q_L; \theta_0) - F'(q_U; \theta_0) = 0$ is unique we have $dP(\theta; \theta_0)/d\theta < 0$ for $\theta < \theta_0$ and $dP(\theta; \theta_0)/d\theta > 0$ for $\theta > \theta_0$ because $H > 0$ for $\theta < \theta_0$ and $H < 0$ for $\theta > \theta_0$. Consistency of the test follows from property 4.

(c) We prove that if any other function $R(S)$, such that $R' > 0$, is used as a test statistic then the test is equivalent to the test based on S. Since R is a strictly increasing function the inverse function, $r = R^{-1}$, exists with the cdf $F_R(r(S); \theta)$. As follows from (3.1), the R-based limits are $q_L = r(q_L)$ and $q_L = r(q_U)$, with the acceptance rule $F(r(q_L); \theta_0) < F(r(S); \theta_0) < F(r(q_U); \theta_0)$. But since r is a strictly increasing function, this rule is equivalent to the acceptance rule based on statistic S. □

The unbiased test and Theorem 3.1 hold for positive argument s and/or parameter θ (formally, 0 is treated as $-\infty$). The following examples illustrate the unbiased test and Theorem 3.1.

Example 3.2 *Unbiased test for normal variance.*

The cdf and pdf of S are $F(s; \sigma^2) = C_{n-1}(s/\sigma^2)$ and $f(s; \sigma^2) = \sigma^{-2}c_{n-1}(s/\sigma^2)$, respectively. Since $F'(s; \sigma^2) = -s/\sigma^4 c_{n-1}(s/\sigma^2)$, system (3.1) turns into

$$C_{n-1}(\sigma_U^2/\sigma_0^2) - C_{n-1}(\sigma_L^2/\sigma_0^2) - (1 - \alpha) = 0,$$
$$(\sigma_L^2/\sigma_0^2)c_{n-1}(\sigma_L^2/\sigma_0^2) - (\sigma_U^2/\sigma_0^2)c_{n-1}(\sigma_U^2/\sigma_0^2) = 0.$$

Letting $q_U = \sigma_U^2/\sigma_0^2$ and $q_L = \sigma_L^2/\sigma_0^2$, we arrive at the familiar system

$$C_{n-1}(q_U) - C_{n-1}(q_L) - (1 - \alpha) = 0,$$
$$(n - 1)(\ln q_U - \ln q_L) - (q_U - q_L) = 0,$$

which is numerically solved by Newton's algorithm realized in the R function `var.ql` with `adj=1`. This solution leads to the unbiased test for the normal variance from Section 1.3 with the power function (1.12). From formula (1.4), we obtain

$$\frac{\partial \ln f(s; \sigma_0^2)}{\partial s} = \frac{n-3}{2s} \ln s - \frac{1}{2\sigma_0^2},$$

and

$$\frac{\partial^2 \ln f(s; \sigma_0^2)}{\partial s \partial \theta} = \frac{1}{2\sigma_0^4} > 0.$$

Condition (3.2) holds, and from Theorem 3.1, one concludes that system (1.11) has a unique solution for any σ_0^2 and $\alpha \in (0, 1)$. The power function has zero derivative at σ_0^2. It is decreasing to the left of σ_0^2 and increasing to the right of σ_0^2, and approaches to 1 when the alternative σ^2 goes to 0 or ∞. Thus, the unbiased test is consistent.

Example 3.3 *Statistical testing for the exponential distribution.*

Given n iid observations from an exponential distribution with the scale parameter θ we want to construct an unbiased test for the null hypothesis $H_0 : \theta = \theta_0$ versus $H_A : \theta \neq \theta_0$ using statistic $S = \sum_{i=1}^{n} Y_i$, which has a chi-square distribution with cdf $F(s; \theta) = C_{2n}(2s/\theta)$ as discussed in Example 2.9. To derive the unbiased test (3.1), similar to the normal variance example, we need to find q_L and q_U such that $C_{2n}(2q_U/\theta_0) - C_{2n}(2q_L/\theta_0) = 1 - \alpha$ and $\theta_0 n \ln(q_U/q_L) - (q_U - q_L) = 0$. Iterations are required to solve this system.

The adjustment vector for (q_L, q_U) in Newton's algorithm takes the form

$$\begin{bmatrix} -2/\theta_0 c_{2n}(2q_L/\theta_0) & 2/\theta_0 c_{2n}(2q_L/\theta_0) \\ -\theta_0 n/q_L + 1 & \theta_0 n/q_U - 1 \end{bmatrix}^{-1}$$

$$\times \begin{bmatrix} C_{2n}(2q_U/\theta_0) - C_{2n}(2q_L/\theta_0) - (1 - \alpha) \\ \theta_0 n \ln(q_U/q_L) - (q_U - q_L) \end{bmatrix}.$$

Typically, it requires three-four iterations to converge starting from the equal-tail quantiles: $q_L = \theta_0 C_{2n}^{-1}(\alpha/2)/2$ and $q_U = \theta_0 C_{2n}^{-1}(1 - \alpha/2)/2$. It is elementary to show that $d^2 \ln f(s; \theta)/ds^2 < 0$ where $f(s; \theta) = (2/\theta) c_{2n}(2s/\theta)$. From Theorem 3.1 it follows that the solution exists and is unique for every θ_0 and $0 < \lambda < 1$. Figure 3.1 depicts the power functions (2.11) for the biased equal-tail and unbiased unequal-tail test with $n = 2$ and $\theta_0 = 0.5$. This graph was created by calling the R function expTCI(job=3). Two-sample testing for the exponential distribution is discussed in Section 5.5.

Example 3.4 *Statistical testing for the Cauchy distribution.*

Statistical testing for the Cauchy distribution is mostly developed based on the maximum likelihood theory (Johnson et al. 1994). We aim to derive an unbiased double-sided test for the scale parameter, $\theta > 0$, in the Cauchy

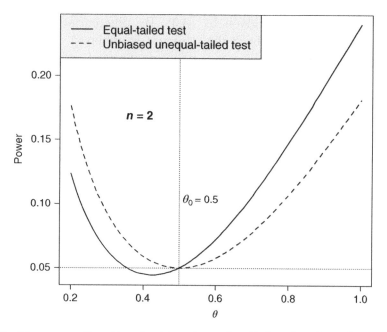

Figure 3.1: *The power function for biased equal-tailed and unbiased unequal-tailed tests around $\theta_0 = 0.5$ with two exponentially distributed observations (n = 2, $\alpha = 0.05$). This graph was created by calling the R function* expTCI(job=3).

distribution (the location parameter is zero). Since the average over a random sample has the same distribution we chose $S = |\overline{X}|$ with the cdf

$$F(s; \theta) = 2/\pi \arctan(s/\theta), \quad s \geq 0.$$

The unbiased test requires solving the system of two equations (3.1), which in our case, turns into

$$\arctan \frac{q_U}{\theta_0} - \arctan \frac{q_L}{\theta_0} = \pi(1 - \alpha)/2,$$

$$\frac{q_L}{\theta_0^2} \frac{1}{1 + (q_L/\theta_0)^2} - \frac{q_U}{\theta_0^2} \frac{1}{1 + (q_U/\theta_0)^2} = 0.$$

The last equation implies $q_L/\theta_0 = 1/(q_U/\theta_0)$, and therefore the first equation turns into $\arctan(q_U/\theta_0) - \arctan(\theta_0/q_U) = \pi(1 - \alpha)$ with a closed-form solution

$$q_U = \theta_0 \tan \left(R + \arctan \left(\sqrt{\tan^2 R + 1} - \tan R \right) \right), \qquad (3.3)$$

where $R = \pi(1 - \alpha)/2$. The null hypothesis $H_0 : \theta = \theta_0$ is rejected if $|X|$ falls outside of the interval $(\theta_0^2/q_U, q_U)$. The power function takes the form

$$P(\theta; \theta_0) = 1 - \frac{2}{\pi} \arctan \frac{\theta}{q_L} + \frac{2}{\pi} \arctan \frac{\theta}{q_U},$$

where $q_L = \theta_0^2/q_U$ and q_U is given by (3.3). By differentiation of this power function, it is elementary to check that this test is unbiased, and by evaluating $P(\theta_0; \theta_0)$, one concludes that the test has the exact type I error α.

3.2 Unbiased CI and MO estimator

The unbiased CI is derived from the unbiased test discussed in the previous section following the principle of duality: the unbiased CI is composed of all θ_0 for which the null hypothesis is accepted. Following this guideline, equations (3.1) the lower and the upper limits of the unbiased CI (θ_L, θ_U), are found by solving the system of equations

$$F(S; \theta_L) - F(S; \theta_U) = \lambda, \ f(S; \theta_L) - f(S; \theta_U) = 0. \tag{3.4}$$

Note that the derivative of the cdf with respect to the parameter changes to the derivative of the cdf with respect to the argument yielding the pdf.

The following result, as a mirror of Theorem 3.1, proves that under certain conditions the solution of system (3.4) is well-defined.

Theorem 3.5 *If in addition to properties 1–4 from Section 2.1, $d^2 \ln f(S; \theta)/d\theta^2 < 0$ for every θ, system (3.4) has a unique solution for every $0 < \lambda < 1$.*

Proof. Similarly to the proof of Theorem 3.1, introduce function

$$H(\theta) \stackrel{\text{def}}{=} f(\theta) - f(F^{-1}(F(\theta) + \lambda)), \quad \theta < F^{-1}(\alpha),$$

where $\theta = \theta_L$ and S is omitted for brevity, and F^{-1} is inverse with respect to the parameter. We have

$$\lim_{\theta \to -\infty} H(\theta) = 0 - f(F^{-1}(\lambda)) < 0,$$

$$\lim_{\theta \to F^{-1}(\alpha)} H(\theta) = f(F^{-1}(\alpha)) - 0 > 0.$$

This proves the existence. Using the chain rule and differentiation of the inverse function, we obtain

$$\frac{dH}{d\theta} = f'(\theta) - \frac{f'(\theta_*)}{f(\theta_*)} f(\theta) = f(\theta) \left(\frac{d}{d\theta} \ln f(\theta) - \frac{d}{d\theta} \ln f(\theta_*) \right),$$

where $\theta_* = F^{-1}(F(\theta) + \lambda) > \theta$. From the condition of the theorem $d \ln f(\theta)/d\theta$ is a decreasing function and therefore $dH/d\theta > 0$. This implies that the solution of system (3.4) is unique. $\square$

Theorem 3.6 *Under assumptions listed in Section 2.1 and conditions listed in Theorem 2.3, the CI defined by system (3.4) is exact and unbiased.*

Proof. The following is a generalization of the unbiased CI for normal variance from Section 1.3.1. The proof closely follows the proof of Theorem 2.3. Remember that for an unbiased CI the probability of coverage of the "wrong" parameter is smaller than the probability of coverage of the true parameter (Rao 1973, p. 470):

$$\Pr(\theta_L(S) \leq \theta \leq \theta_U(S)) < \Pr(\theta_L(S) \leq \theta_0 \leq \theta_U(S)), \ \theta \neq \theta_0, \tag{3.5}$$

where $\theta_L = \theta_L(S)$ and $\theta_U = \theta_U(S)$ are strictly increasing functions of S implied by the assumptions. Here, θ is treated as the "wrong" and θ_0 as the true parameter. Express the coverage probability as

$$\Pr(\theta_L(S) \leq \theta \leq \theta_U(S)) = \Pr(\theta_L(S) \leq \theta) - \Pr(\theta_U(S) \leq \theta)$$

$$= \Pr(S \leq \theta_L^{-1}(\theta)) - \Pr(S \leq \theta_U^{-1}(\theta))$$

$$= F(\theta_L^{-1}(\theta); \theta_0) - F(\theta_U^{-1}(\theta); \theta_0),$$

where θ_L^{-1} and θ_U^{-1} are the inverse functions. A necessary condition for (3.5) to hold is that the derivative of the left-hand side with respect to θ at θ_0 equals zero for every θ_0. The derivative of the right-hand side of the last expression evaluated at θ_0 gives

$$\frac{d}{d\theta} \Pr(\theta_L(S) \leq \theta \leq \theta_U(S))|_{\theta=\theta_0}$$

$$= (\theta_L^{-1}(\theta_0))' f(\theta_L^{-1}(\theta_0); \theta_0) - (\theta_U^{-1}(\theta_0))' f(\theta_U^{-1}(\theta_0); \theta_0) = 0,$$

where $'$ means the derivative of the inverse function. Now make a change of variable in the first term: $\theta_L^{-1}(\theta_0) = s$, which implies $\theta_0 = \theta_L(s)$ and $(\theta_L^{-1}(\theta_0))' = 1$ because the derivative of the inverse function is reciprocal of the original function. The same transformation applies to the second term so that the necessary condition in terms of s is rewritten as $f(s; \theta_L(s)) - f(s; \theta_U(s)) = 0$ for every s, which is equivalent to the second equation of (3.4) at the observed $s = S$. $\square$

System (3.4) can be effectively solved by Newton's algorithm with the adjustment vector for (θ_L, θ_U) given by

$$\begin{bmatrix} F'(S; \theta_L) & -F'(S; \theta_U) \\ f'(S; \theta_L) & -f'(S; \theta_U) \end{bmatrix}^{-1} \begin{bmatrix} F(S; \theta_L) - F(S; \theta_U) - \lambda \\ f(S; \theta_L) - f(S; \theta_U) \end{bmatrix}$$

starting from the equal-tail confidence limits, where differentiation is understood with respect to the parameter, as previously.

Example 3.7 *Unbiased CI for normal variance.*

We apply the CI (3.4) to the normal variance with $f(s; \sigma^2) = \sigma^{-2}c_{n-1}(s\sigma^{-2})$. The second equation turns into

$$(S\sigma_L^{-2})c_{n-1}(S\sigma_L^{-2}) - (S\sigma_U^{-2})c_{n-1}(S\sigma_U^{-2}) = 0,$$

which is equivalent to (1.10), where $q_L = S\sigma_L^{-2}$ and $q_U = S\sigma_U^{-2}$. $\qquad\square$

Now we are ready to define the MO estimator as the limit point of the unbiased CI (θ_L, θ_U) defined from (3.4).

Definition 3.8 *The mode (MO) estimator $\widehat{\theta}_{MO} = \widehat{\theta}_{MO}(S)$ is the limit point of the unbiased CI when the confidence level approaches zero, $\lambda \to 0$.*

When $\lambda \to 0$ from the first equation of (3.4) we conclude that $\theta_L \to \theta$ and $\theta_U \to \theta$. From the second equation we derive

$$0 = \lim_{\lambda \to 0} \frac{f(S; \theta_U) - f(S; \theta_L)}{\theta_U - \theta_L} = f'(S; \theta).$$

This means the MO estimator is the solution of the equation

$$\left.\frac{\partial f(S; \theta)}{\partial \theta}\right|_{\theta=\widehat{\theta}_{MO}} = 0. \qquad (3.6)$$

In other words, the MO estimator is the value of the parameter that maximizes the pdf given the observed statistic S. There is a similarity between the maximum likelihood (ML) and MO estimator. The difference is that here S is a univariate statistic as a function of data, not the data themselves. The MO estimator shares the same invariance property with the ML: the MO estimator of $\tau = \tau(\theta)$ is $\tau(\widehat{\theta}_{MO})$, where τ is a monotonic transformation. In short, reparametrization does not change the MO theory – the invariance property of the MO inference is illustrated in Section 5.3 and further employed to standard deviation in Section 5.1 and the coefficient of variation in Section 5.6.2. Note that while the ML inference is supported by the optimal inference in the large sample the MO inference is exact even for an extremely small sample size.

Theorem 3.9 *Under the conditions of Theorem 3.5, the MO estimator exists and is unique.*

Proof. From elementary calculus, it follows that $f(S; \theta) \to 0$ when $\theta \to \infty$ or $\theta \to -\infty$ because $F(S; \theta) \to 0$ or $F(S; \theta) \to 1$, respectively. As in the maximum likelihood theory, introduce the log pdf function $l(\theta) = \ln f(S; \theta)$ for which $l(\theta) \to \pm\infty$ when $\theta \to \pm\infty$. Obviously, the solution to the equation (3.6) exists and is unique if and only if the same statement holds for function $l(\theta)$. Condition $d^2 \ln f(S; \theta)/d\theta^2 < 0$ guarantees that function $l(\theta)$ is strictly concave and since $l(\theta) \to \pm\infty$ it has a unique stationary point, the MO estimator. $\qquad\square$

The following result connects ML and MO estimators for a special case.

Theorem 3.10 *If S is the sufficient statistic then the MO and ML estimators coincide.*

Proof. If S is the sufficient statistic, then the pdf of $\mathbf{y}$ can be expressed as $q(\mathbf{y}; \theta) = g(S(\mathbf{y}); \theta)h(\mathbf{y})$. Define the transformation

$$s = S(\mathbf{y}), u_2 = y_2, ..., u_n = y_n.$$

Assume that function S has a unique inverse for y_k for some k given the rest $y_i, i \neq k$. Without loss of generality we can assume that $k = 1$. Denote this inverse function as $y_1 = w(s, y_2, ..., y_n)$. Then upon transformation

$$s = S(y_1, y_2, ..., y_n), \ y_2 = u_2, ..., y_n = u_n$$

the density $q(\mathbf{y}; \theta)$ can be written as $g(s; \theta)r(s, u_2, ..., u_n)$, where

$$r(s, u_2, ..., u_n) = \left| \frac{\partial w(s, u_2, ..., u_n)}{\partial s} \right| h(w(s, u_2, ..., u_n), u_2, ..., u_n).$$

Finally, the density of the sufficient statistic S can be expressed as

$$f(s; \theta) = g(s; \theta) \int_{R^{n-1}} r(s, u_2, ..., u_n) du_2 \cdots du_n.$$

The maximum of $f(S; \theta)$ over θ is attained at the same parameter value as the ML estimate. □

Example 3.11 *The MO and ML estimators for the normal variance when μ is known.*

The ML estimator of the normal variance is $\sigma^2_{ML} = \sqrt{S_0/n}$, where $S_0 = \sum(Y_i - \mu)^2$. Since $f(s; \sigma^2) = \sigma^{-2}c_n(s\sigma^{-2})$, we have

$$\ln f(S_0; \sigma^2) = -\ln \sigma^2 + \frac{n-2}{2} \ln \frac{S_0}{\sigma^2} - \frac{S_0}{2\sigma^2}$$

$$= \frac{n-2}{2} \ln S - \frac{n}{2} \ln \sigma^2 - \frac{S_0}{2\sigma^2}.$$

Upon differentiation with respect to σ^2, we obtain that the maximum of $f(S_0; \sigma^2)$ is attained at $\sigma^2_{MO} = \sigma^2_{ML}$.

3.3 Cumulative information and the sufficient statistic

The goal of this section is to introduce the efficiency criterion for the MO estimator and show that a sufficient statistic gives birth to the most efficient estimator.

As in the case of the MC estimator, estimators are compared on the cdf scale: that is, the efficiencies of estimators are compared under the condition that they have the same cdf value (see Section 2.4). Our efficiency criterion has much in common with the Fisher information (Rao 1973) but is more specific with respect to the value of the statistic. We show that the efficient MO estimator yields maximum Fisher information.

Definition 3.12 *Let random vector* $\mathbf{Y}$ *have pdf* $f(\mathbf{y}; \theta)$ *dependent on parameter* θ. *Define information element as*

$$i(\mathbf{y}; \theta) = -\frac{d^2 \ln f(\mathbf{y}; \theta)}{d^2 \theta}.$$

Define cumulative information function *for statistic* $Q = Q(\mathbf{Y})$ *as*

$$I_Q(s; \theta) = \int_{Q(\mathbf{y}) \le s} i(\mathbf{y}; \theta) f(\mathbf{y}; \theta) d\mathbf{y}, \; -\infty < s < \infty. \tag{3.7}$$

When $s \to \infty$ cumulative information approaches the traditional Fisher information. When the support of Q is positive values we use $0 < s < \infty$.

Definition 3.13 *We say that statistic* $S = S(\mathbf{y})$ *is no less informative than statistic* $R = R(\mathbf{y})$ *for given* θ *if* $I_S(s_1; \theta) \ge I_R(s_2; \theta)$ *for every* s_1, s_2 *such that* $F_S(s_1; \theta) = F_R(s_2; \theta)$. *If this holds for every* θ, *we say that* S *is* uniformly *no less informative than* R.

Our definition is more stringent than the classic Fisher information comparison because the former turns into the latter when $s_1, s_2 \to \infty$. The inequality of $I_S(s_1; \theta) \ge I_R(s_2; \theta)$ under condition $F_S(s_1; \theta) = F_R(s_2; \theta)$ can be tested on the scale of $0 < p < 1$ by defining $s_1 = Q_S(p; \theta)$ and $s_2 = Q_R(p; \theta)$, where Q is the quantile function. Then the inequality $I_S(s_1; \theta) \ge I_R(s_2; \theta)$ can be equivalently written as $I_S(p; \theta) \ge I_R(p; \theta)$ for every $p \in (0, 1)$.

Theorem 3.14 *Let* $S = S(\mathbf{y})$ *be a sufficient statistic, that is, the pdf of* $\mathbf{y}$ *can be expressed as* $f(\mathbf{y}; \theta) = e^{g(S(\mathbf{y}); \theta) + p(\theta) + h(\mathbf{y})}$. *Suppose that* $g''_{\theta\theta}(s; \theta) = \partial^2 g / \partial \theta^2$ *is a strictly decreasing function of* s *and* $R = R(\mathbf{y})$ *is another statistic such that* $F_S(s_1; \theta) = F_R(s_2; \theta)$ *for some* s_1 *and* s_2. *Then*

$$I_S(s_1; \theta) \ge I_R(s_2; \theta), \tag{3.8}$$

that is, the sufficient statistic yields a MO estimator no less efficient than any other statistic for any given θ.

Proof. In the integral form, condition $F_S(s_1; \theta) = F_R(s_2; \theta)$ is rewritten as

$$\int_{S(\mathbf{y}) \le s_1} f(\mathbf{y}; \theta) d\mathbf{y} = \int_{R(\mathbf{y}) \le s_2} f(\mathbf{y}; \theta) d\mathbf{y}. \tag{3.9}$$

To prove (3.8) it suffices to show that

$$-\int_{R(\mathbf{y}) \leq s_2} (g_{\theta\theta}''(S(\mathbf{y});\theta) + p''(\theta)) f(\mathbf{y};\theta) d\mathbf{y}$$

$$\leq -\int_{S(\mathbf{y}) \leq s_1} (g_{\theta\theta}''(S(\mathbf{y});\theta) + p''(\theta)) f(\mathbf{y};\theta) d\mathbf{y},$$

which, due to (3.9), is equivalent to

$$\int_{R(\mathbf{y}) \leq s_2} - g_{\theta\theta}''(S(\mathbf{y});\theta) d\mathbf{y} \leq \int_{S(\mathbf{y}) \leq s_1} - g_{\theta\theta}''(S(\mathbf{y});\theta) f(\mathbf{y};\theta) d\mathbf{y}.$$

The rest of the proof follows the proof of Theorem 2.16 with $g_\theta'(S(\mathbf{y});\theta)$ replaced with an increasing function $-g_{\theta\theta}''(S(\mathbf{y});\theta)$. □

Schervish (Theorem 2.86, p. 113) proves that the Fisher information in any statistic $R = R(\mathbf{Y})$ is no greater than in the whole sample, and if R is the sufficient statistic ($S = R$), the two information quantities are the same. This makes people saying that sufficient statistic possesses all information about the sample. The cumulative information function along with Definition 3.13 and Theorem 3.14 establishes a stringent ordering between statistics having the same cdf value. As in the traditional setting, Theorem 3.14 proves that the sufficient statistic is the most informative.

The following example illustrates the superiority of the sufficient statistic for normal variance with known mean, similar to Example 2.17 from Section 2.4.

Example 3.15 *Efficiency for normal variance.*

It is easy to see that the condition of Theorem 3.14 is fulfilled because $g''(s;\sigma^2) = -s/\sigma^6$ is a decreasing function of s. As in Example 1, we compare two statistics S and R. Find $I_S(s;\sigma^2)$ following Definition (3.7). Omitting some algebra, we derive

$$I_S(s;\sigma^2) = n\sigma^{-4} C_{n+2}(s/\sigma^2) - 0.5n\sigma^{-2} C_n(s/\sigma^2),$$
$$I_R(s;\sigma^2) = (n-1)\sigma^{-4} C_{n+1}(s/\sigma^2) - 0.5(n-1)\sigma^{-2} C_{n-1}(s/\sigma^2).$$

Figure 3.2 depicts cumulative information functions I_S and I_R on the probability scale as the respective cdf values (p) with the true $\sigma^2 = 0.3$ for two sample sizes. As we can see the I_S curve is above I_R, as expected from Theorem 3.14. When s is large the information criterion approaches the classic (cumulative) Fisher information shown as the circle at $p = 1$. □

The following examples illustrate the MO exact statistical inference.

Example 3.16 *The MO estimator for normal variance.*

Since $f(s;\theta) = \sigma^{-2} c_{n-1}(\sigma^{-2}s)$, the second equation of (3.4) turns into $(n-1)\ln(\sigma_L^2/\sigma_U^2) - S(\sigma_L^{-2} - \sigma_U^{-2}) = 0$. Letting $q_L = S\sigma_L^{-2}$ and $q_U = S\sigma_U^{-2}$, we

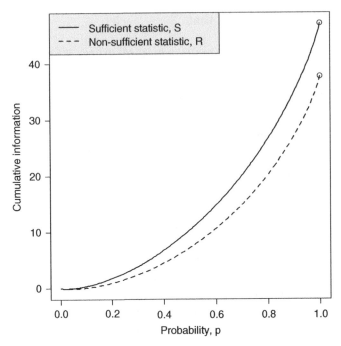

Figure 3.2: *Comparison of two MO estimators for normal variance using cumulative information functions (Example 3.15) with $n = 5$. The S-derived MO estimator of σ^2 is more efficient than the R-derived (the circle indicates Fisher information).*

rewrite the previous equation as $(n - 1) \ln(q_U/q_L) - (q_U - q_L) = 0$, which is the same as (1.11). Now we illustrate the MO estimator using formula (3.6). Since the pdf of S is $\sigma^{-2} c_{n-1}(s\sigma^{-2})$, we have

$$\left. \frac{\partial \ln(\sigma^{-2} c_{n-1}(s\sigma^{-2}))}{\partial s} \right|_{s=S} = \frac{n-3}{2S} - \frac{1}{2\sigma^2} = 0$$

with solution

$$\widehat{\sigma}^2_{MO} = \frac{1}{n-3} \sum_{i=1}^{n} (Y_i - \overline{Y})^2.$$

Example 3.17 *Scale/rate parameter.*

This example generalizes statistical inference for normal variance and an exponential distribution by assuming that the cdf of statistic S takes the form $F(s; \theta) = H(s/\theta)$ when θ is the scale parameter, and $F(s; \lambda) = H(\lambda s)$ when λ is the rate parameter, both positive, with known cdf H. Also, we assume that the density function, $h = H'$ is unimodal with mode M. Elementary calculus yields the MO estimators $\widehat{\theta}_{MO} = S \times M$ and $\widehat{\lambda}_{MO} = S/M$.   □

More examples of MO statistical inference applied to popular statistical parameters are found in the next chapter.

References

[1] Casella, G. and Berger, R.L. (1990). *Statistical Inference*. Belmont, CA: Duxbury Press.

[2] Johnson, N.L., Kotz, S., and Balakrishnan, N. (1994). *Continuous Univariate Distributions*, vol. 1, 2e. New York: Wiley.

[3] Lehmann, E.L. and Romano, J.P. (2005). *Testing Statistical Hypothesis*, 3e. New York: Springer.

[4] Rao, C.R. (1973). *Linear Statistical Inference and Its Applications*, 2e. New York: Wiley.

[5] Schervish, M. (2012). *Theory of Statistics*. New York: Springer.

[6] Shao, J. (2003). *Mathematical Statistics*. New York: Springer.

Chapter 4

P-value and duality

First, we introduce a new definition of the p-value for the double-sided hypothesis, which is especially important for asymmetric distributions. This definition implies that under the null hypothesis, the distribution of the p-value is uniform on $(0, 1)$, and under the alternative yields the power functions defined in the previous chapters: equal-tail, DL, and unbiased test. Second, following the duality principle between CI and hypothesis test, we derive specific forms of the test dual to CI and CI dual to test for the MC and MO-statistics. Finally, the M-statistics for exact statistical inference is overviewed in the last section, accompanied by the table of major equations.

4.1 P-value for the double-sided hypothesis

There are two paradigms of statistical hypothesis testing, known by the names of the great statisticians, Fisher, Neyman and Pearson. While the latter operates with the power function the former solely relies on the computation of the p-value. The two schools of thought fought throughout the previous century. While the Fisherian approach was mostly adopted by applied statisticians, due to its simplicity, the Neyman & Pearson approach occupies the mind of theoreticians. Neither of the two approaches yields a definite answer on hypothesis testing: in the Fisherian approach, a threshold for the p-value must be specified, such as 0.05, and in the Neyman & Pearson approach, in addition to knowing α, the alternative is assumed to be known. Using the universal threshold $\alpha = 0.05$ for statistical significance throughout projects with different goals in mind and across disciplines creates many problems, and the recent unprecedented statement of the American Statistical Association reflects this fact (Wasserstein and Lazar 2016).

So far, our hypothesis testing has relied on the power function as the centerpiece of the Neyman & Pearson approach. In this section, we discuss an alternative Fisherian approach based on the computation of the *p*-value. The verbal definition of the *p*-value, as the probability of obtaining results at least as extreme as the observed results of a statistical hypothesis test under the null hypothesis, is straightforward to apply to a one-sided hypothesis or symmetric distribution. However, to date, there is no consensus on how to translate "at least as extreme" into a rigorous rule for a two-sided hypothesis with the asymmetric test statistic. Even Fisher himself (1925) suggested doubling the one-sided *p*-value to obtain the two-sided *p*-value without proof (we prove that this suggestion is true when the equal-tail probability test is used). As Dunne et al. (1996) put it: "The lack of thoughtful discussion of this problem in most statistical texts is rather surprising."

The goal of this section is to propose a new definition of the *p*-value for two-sided hypotheses in the framework of M-statistics. The definition of the *p*-value is the same for MC and MO-statistics but the values are different because they use different q_L and q_U. Our definition marries Fisher and Neyman & Pearson because we prove that under the null hypothesis, our p has the uniform distribution on $(0, 1)$ and under the alternative the probability that p is smaller than the significance level, α, yields the power of the test as defined in the previous chapter.

4.1.1   General definition

We are concerned with testing the double-sided null hypothesis $H_0 : \theta = \theta_0$ versus the alternative $H_A : \theta \neq \theta_0$ having the test statistic S with cdf $F(s; \theta)$. As follows from our previous discussion, our test requires computing quantiles q_L and q_L such that

$$F(q_U; \theta_0) - F(q_L; \theta_0) = 1 - \alpha. \qquad (4.1)$$

Three methods for computing the quantiles have been discussed. In the traditional equal-tail test θ_L and θ_U, we computed as solutions of the equations $F(\theta_L; \theta_0) = \alpha/2$ and $F(\theta_U; \theta_0) = 1 - \alpha/2$. The DL and unbiased tests require the solution of the coupled system of equations, (2.8) and (3.1), respectively. At this point, we do not care how quantiles are computed, but they must satisfy (4.1). For all these cases the power function is given by

$$P(\theta; \theta_0) = 1 + F(q_L; \theta) - F(q_U; \theta), \qquad (4.2)$$

where θ is the alternative parameter value (quantiles q_L and q_U depend on θ_0 only).

If S is the observed value, the general idea for computing the p-value follows the traditional recipe: on the left of S, the p-value is proportional to $F(S; \theta_0)$, the left tail probability, and on the right of S is proportional to $1 - F(S; \theta_0)$, the right tail probability. The major question is: left and right of what? In other words, what is the value of S that serves as the divider? Classic statistics does not answer. In our definition, below the divider value is where $F(S; \theta_0) = F(\theta_L; \theta_0)/\alpha$.

Definition 4.1 *Define the p-value for the two-sided test with significance level α as*

$$p = \begin{cases} \frac{\alpha}{F(q_L; \theta_0)} F(S; \theta_0) & \text{if } F(S; \theta_0) \leq \frac{1}{\alpha} F(q_L; \theta_0) \\ \frac{\alpha}{\alpha - F(q_L; \theta_0)} (1 - F(S; \theta_0)) & \text{if } F(S; \theta_0) > \frac{1}{\alpha} F(q_L; \theta_0) \end{cases} \tag{4.3}$$

where the quantiles q_L and q_U satisfy (4.1).

This definition applies to the equal-tail, unbiased, and DL tests but with different q_L and q_U. Under this definition, p depends on the nominal significance level, α, and as such, is sometimes referred to as the α-dependent p-value.

Alternatively, the p-value may be computed as

$$p = \min \left(\frac{\alpha F(S; \theta_0)}{F(q_L; \theta_0)}, \frac{\alpha(1 - F(S; \theta_0))}{\alpha - F(q_L; \theta_0)} \right). \tag{4.4}$$

To prove, denote $A = F(S; \theta_0)$ and $B = F(q_L; \theta_0)$, where $0 < A < 1$ and $0 < B < \alpha$. Elementary algebra proves that if $A \leq B/\alpha$ then $\alpha A/B \leq \alpha(1 - A)/(\alpha - B)$. Now consider the case when $A > B/\alpha$. Again, elementary algebra proves that $\alpha/(\alpha - B)(1 - A) < \alpha/B$. Thus, p is the minimum of B/α and $\alpha/(\alpha - B)(1 - A)$.

Lemma 4.2 *If $X \sim \mathcal{R}(0, 1)$ and b is a fixed number, $b \in (0, 1)$, then the random variable*

$$Y = \begin{cases} \frac{X}{b} & \text{if } X \leq b \\ \frac{1 - X}{1 - b} & \text{if } X > b \end{cases}$$

has also a uniform distribution $\mathcal{R}(0, 1)$.

Proof. Let y be a fixed number such that $0 < y < 1$. Then the cdf of Y is

$$\Pr(Y \leq y) = \Pr(Y \leq y \cap X \leq b) + \Pr(Y \leq y \cap X > b)$$
$$= \Pr\left(\frac{X}{b} \leq y \cap X \leq b \right) + \Pr\left(\frac{1 - X}{1 - b} \leq y \cap X > b \right)$$
$$= \Pr(X \leq by \cap X \leq b) + \Pr(X \geq 1 - (1 - b)y \cap X > b).$$

Since $by \leq y$ and $1 - (1-b)y \geq b$ we have

$$\Pr(Y \leq y) = \Pr(X \leq by) + \Pr(X \geq 1 - (1-b)y)$$
$$= by + 1 - (1 - (1-b)y) = y.$$

That is, the cdf of Y is y on $(0,1)$.   □

Theorem 4.3 *Properties of the p-value. (a) The p-value defined by (4.3) has a uniform distribution on $(0,1)$ under the null hypothesis. (b) Under the alternative hypothesis this p-value yields the power function (4.2), that is,*

$$\Pr(p \leq \alpha; \theta) = 1 + F(q_L; \theta) - F(q_U; \theta). \tag{4.5}$$

Proof. (a) The uniformity follows from the lemma after recognizing that under the null hypothesis $X = F(S; \theta_0) \sim \mathcal{R}(0,1)$. Letting $b = F(q_L; \theta_0)/\alpha$, we deduce that $Y = p$, and therefore the lemma applies. (b) Denote $b = F(q_L; \theta_0)/\alpha$, a fixed constant, where $X = F(S; \theta_0)$ is a random variable that takes values on $(0,1)$, but now X does not follow the uniform distribution on $(0,1)$ because the cdf of S is different from $F(\cdot; \theta_0)$. The cdf of X is

$$\Pr(X \leq x) = \Pr(F(S; \theta_0) \leq x) = F(F^{-1}(x; \theta_0); \theta),$$

where F^{-1} is the inverse cdf at θ_0, the quantile function. Following the proof of Lemma 4.2, we obtain

$$\Pr(p \leq \alpha; \theta) = \Pr(F(S; \theta_0) \leq b\alpha; \theta)$$
$$+ 1 - \Pr(F(S; \theta_0) < 1 - (1-b)\alpha; \theta)$$
$$= F(F^{-1}(b\alpha; \theta_0); \theta) + 1 - F(F^{-1}(1 - (1-b)\alpha; \theta_0); \theta)$$
$$= F(q_L; \theta) + 1 - F(q_U; \theta) = P(\theta; \theta_0),$$

the power function defined in (4.2).   □

Corollary 4.4 *The p-value for the equal-tail test.*

In the equal-tail test, quantiles are found from $F(q_L; \theta_0) = \alpha/2$ and $F(q_U; \theta_0) = 1 - \alpha/2$. Therefore (4.3) turns into

$$p = \begin{cases} 2F(S; \theta_0) \text{ if } F(S; \theta_0) \leq 0.5 \\ 2(1 - F(S; \theta_0)) \text{ if } F(S; \theta_0) > 0.5 \end{cases}$$

or in the language of the median

$$p = \begin{cases} 2F(S; \theta_0) \text{ if } S \leq \texttt{median} \\ 2(1 - F(S; \theta_0)) \text{ if } S > \texttt{median} \end{cases}$$

where `median` is understood as the median of the distribution at the null. Equivalently, as follows from (4.4), the p-value of the equal-tail test can be computed as

$$p = 2 \times \min(F(S; \theta_0), 1 - F(S; \theta_0)), \tag{4.6}$$

the double least tail probability.

Corollary 4.5 *An alternative definition of the p-value for the DL test.*

Definition 4.1 appeals to the cdf for computing "the probability of obtaining results at least as extreme as the observed results." For the DL test, it is natural to take the pdf as a metric for computation of the probability that future observations have pdf values greater than $f(S; \theta_0)$. This idea leads to the following alternative definition of the p-value for the DL test:

$$p = 1 - \int_{\{f(s;\theta_0) \geq f(S;\theta_0)\}} f(s; \theta_0) ds, \tag{4.7}$$

where S is treated as fixed. Since f is a unimodal function $\{f(s; \theta_0) \geq f(S; \theta_0)\}$ defines an interval with the limits as the solution of equation $f(s; \theta_0) = f(S; \theta_0)$. An obvious solution is S itself, and the second solution, $\tau = \tau(S)$, defines another limit. Therefore (4.7) is rewritten as

$$p = 1 - |F(\tau; \theta_0) - F(S; \theta_0)| \tag{4.8}$$

where τ is such that

$$f(\tau; \theta_0) = f(S; \theta_0), \quad \tau \neq S. \tag{4.9}$$

Theorem 4.6 *(a) If S has cdf $F(\cdot; \theta_0)$, then p defined by (4.8) and (4.9) has a uniform distribution on $(0, 1)$. (b) If S has cdf $F(\cdot; \theta)$. then (4.5) holds.*

Proof. We merely let $q_L = \min(\tau, S)$, $q_U = \max(\tau, S)$ and then invoke Theorem 4.3. □

The alternative definition of the p-value does not depend on α but has the same properties as the α-dependent p-value for the DL test.

Corollary 4.7 *p-value as a function of S and p-divider.*

As follows from the definition (4.3), the p-value increases with S before reaching $p = 1$ at the point where

$$F(S; \theta_0) = \alpha^{-1} F(q_L; \theta_0). \tag{4.10}$$

To the right of this point, called the p-divider and denoted as q_p, the p-value decreases to 0. For the equal-tail test, as follows from Corollary 4.4, the p-divider:

that is, $p = 1$, where S is the median of the null distribution. On the left side of the median p increases from 0 to 1, and on the right side of the median it decreases from 1 to 0. In the extreme case when $\alpha \to 1$ the p-divider for the DL and unbiased tests are found from

$$\frac{\partial}{\partial s} f(s; \theta_0)\Big|_{s=q_p} = 0, \quad \frac{\partial}{\partial s} F'(s; \theta_0)\Big|_{s=q_p} = \frac{\partial}{\partial \theta} f(q_p; \theta)\Big|_{\theta=\theta_0} = 0,$$

respectively. In other words, when $\alpha \to 1$, the α-dependent p-divider for the DL test is the mode and for the unbiased test is where F has the steepest slope. As follows from (4.7), The alternative p-divider for the DL test is the mode as well. The three tests are illustrated with the normal variance in the next section.

Example 4.8 *P-value for the symmetric test statistic.*

We aim to show that when the test statistic S has a symmetric distribution around θ_0, such as in the t-test, the p-value defined by (4.3) coincides with the traditional p-value by doubling the tail probability. When S has a symmetric distribution its cdf takes the form $F(s; \theta_0) = H(s - \theta_0)$, where $h = H'$ is the symmetric pdf around zero. For example, for the t-test we have $S = \sqrt{n}(\overline{Y} - \theta_0)/\hat{\sigma}_Y$, and H is the cdf of the t-distribution with $n - 1$ df, where $Y_i \overset{iid}{\sim} \mathcal{N}(\theta_0, \sigma^2)$, $i = 1, 2, ..., n$. h being symmetric implies that $F(q_L; \theta_0) = H(q_L - \theta_0) = \alpha/2$ and in the notation of Definition 4.1, we obtain

$$\alpha/F(q_L; \theta_0) = 2, \quad \alpha/(\alpha - F(q_L; \theta_0)) = 2.$$

Therefore the equal-tail test p-value, computed by formula (4.3), turns into

$$p = \begin{cases} 2H(S - \theta_0) \text{ if } S \le \theta_0 \\ 2(1 - H(S - \theta_0)) \text{ if } S > \theta_0 \end{cases} = 2H(-|S - \theta_0|),$$

the traditional p-value by doubling the tail probability.   □

Note that the doubling rule applies only either to an equal-tail test or when the distribution of the test statistic is symmetric.

4.1.2   *P*-value for normal variance

This section illustrates the computation of the p-value for testing a double-sided hypothesis for the normal variance with the null $H_0 : \sigma^2 = \sigma_0^2$ against the alternative $H_A : \sigma^2 \ne \sigma_0^2$.

Equal-tail test

Although there is no clear explanation or proof in the literature, it is suggested to compute the p-value by the formula

$$p = 2 \times \min\left(C_{n-1}(S/\sigma_0^2), 1 - C_{n-1}(S/\sigma_0^2)\right). \tag{4.11}$$

Corollary 4.4 confirms this formula as a special case of Definition 4.1. Equivalently (4.11) can be written as

$$p = \begin{cases} 2C_{n-1}(S/\sigma_0^2) \text{ if } C_{n-1}(S/\sigma_0^2) < 0.5 \\ 2(1 - C_{n-1}(S/\sigma_0^2)) \text{ if } C_{n-1}(S/\sigma_0^2) > 0.5 \end{cases}.$$

As follows from Theorem 4.3, the probability that the p-value is less than α equals the power of the equal-tail test. When $\sigma^2 = \sigma_0^2$, the p-value follows a uniform distribution on $(0, 1)$.

Density level test

See Example 2.14 and equation (2.12) on how to find the quantiles for the DL test. Once q_L and q_U are determined, the p-value is computed as

$$p = \begin{cases} \frac{\alpha}{C_{n-1}(q_L)} C_{n-1}(S/\sigma_0^2) \text{ if } C_{n-1}(S/\sigma_0^2) \leq \frac{C_{n-1}(q_L)}{\alpha} \\ \frac{\alpha}{\alpha - C_{n-1}(q_L)}(1 - C_{n-1}(S/\sigma_0^2)) \text{ if } C_{n-1}(S/\sigma_0^2) > \frac{C_{n-1}(q_L)}{\alpha} \end{cases}. \tag{4.12}$$

Remember that we called this p-value α-dependent (typically, we choose $\alpha = 0.05$). Alternatively, we proposed to compute the p-value by (4.8). This p-value does not depend on α and is different from (4.12) but leads to the same power function. For normal variance, it takes the form

$$p = 1 - |C_{n-1}(q_L/\sigma_0^2) - C_{n-1}(S/\sigma_0^2)|. \tag{4.13}$$

To compute q_L, we need to solve the equation $c_{n-1}(q_L/\sigma_0^2) = c_{n-1}(S/\sigma_0^2)$, which simplifies to

$$\ln s - s/A - B = 0, \quad s > 0, \tag{4.14}$$

where $s = q_L \neq S$, $A = (n-3)\sigma_0^2$, and $B = \ln S - S/A$. Equation (4.14) frequently emerges in connection with normal variance – next we prove that it can be effectively solved by Newton's algorithm with a special choice of starting values.

Proposition 4.9 *The equation (4.14), where $A > 0$ and $B < \ln A - 1$, has two solutions $0 < s_1^* < s_2^*$ found iteratively by Newton's algorithm*

$$s_{k+1} = s_k - \frac{\ln s_k - s_k/A - B}{1/s_k - 1/A}, k = 0, 1, ...,$$

When starting from $s_0 = e^B$, iterations monotonically converge to s_1^*, and when starting from $s_0 = A(\ln A - B)$, they monotonically converge to s_2^*.

Proof. We prove that the equation has exactly two solutions. The existence follows from elementary calculus, which proves that the maximum $\ln A - 1$ of $\ln s - s/A$ is attained at $s = A$. Moreover, the function $\ln s - s/A$ is concave and approaches $-\infty$ when $s \to 0$ or $s \to \infty$. This implies that there are exactly two solutions on both sides of $s = A$. To prove that Newton's algorithm monotonically converges to the solutions we employ the result from Ortega and Rheinboldt (2000) that says Newton's iterations monotonically converge for a concave function if the starting values are to the left and right of the respective solutions. To find these starting values we observe an elementary inequality $\ln s - s/A < \ln s$ that yields $s_0 = e^B$ and $\ln s - s/A > \ln A - s/A$ for $s > A$ that yields $s_0 = A(\ln A - B)$.   □

This algorithm is realized in the R function `lnsAB` – see the R code `pvalVAR`. This function returns the two solutions $s_1^* < s_2^*$ (one of them is associated with S). Note that $B < \ln A - 1$, with A and B defined earlier, follows from the inequality $\ln s - s/A < \ln s$ unless $S = (n-3)\sigma_0^2$, which implies a unique solution.

Figure 4.1 displays four α-dependent p-values for the DL test computed by formula (4.12) against the alternative p-value computed by formula (4.13). As we learn from this plot the α-dependent and alternative p-values are quite different for small α and n.

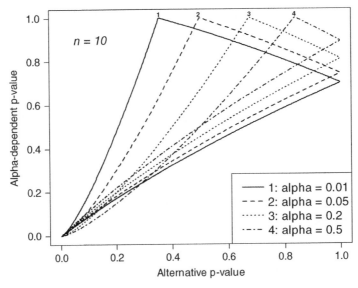

Figure 4.1: *Four α-dependent p-values versus the alternative p-value for the DL test. This graph was created by calling* **pvalVAR_comp(job=1)**.

Unbiased test

The formula for the p-value is the same as given by (4.12), but a different q_L is found by solving (3.1): that is, in the case of variance turns into solving (1.2) and (1.11).

Four p-values and p-dividers

For better interpretation, we display p-values on the scale of the normalized variance, $\eta = \hat{\sigma}^2/\sigma_0^2 = \sigma_0^{-2}S/(n-1)$. Figure 4.2 depicts four p-values as a function of η for $n = 10$ and $\alpha = 0.05$. All functions increase to the left of the p-divider and decrease to the right. The p-divider for the equal-tail (ET) test on the η-scale is $C_{n-1}^{-1}(0.5)/(n-1) < 1$. The p-divider for the α-dependent DL test is $\eta = (n-1)^{-1}C_{n-1}(\alpha^{-1}C_{n-1}(q_L/\sigma_0^2))$. The p-divider of the alternative DL test is the mode or on the η-scale $(n-3)/(n-1)$. The formula for the p-divider for the unbiased test is the same but with different q_L.

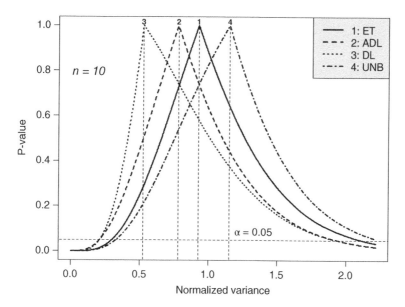

Figure 4.2: *Four p-values as functions of the normalized variance, $\hat{\sigma}^2/\sigma_0^2$ with $n = 10$ and $\alpha = 0.05$. The vertical line at $p = 1$ depicts the p-divider on the η scale (normalized variance). This graph was created by calling* **pvalVAR_comp(job=2)**.

The alternative p-divider for the DL test on the original scale is $\hat{\sigma}_{MC}^2 = S/(n-3)$, as follows from Example 3.16. The p-divider for the unbiased test is $\hat{\sigma}_{MO}^2 = S/(n-1)$. See Section 1.3.

Power function via the *p*-value

The power of the test is the probability that the *p*-value is smaller than the nominal significance level – a connecting property between Fisher and Neyman & Pearson's theories. We test three powers (ET, DL, and UNB) via four *p*-values using simulations (the number of simulations = 1 million) – see R function pval-VAR, which produces Figure 4.3. The theoretical power function was derived in Section 1.3 and has a common expression:

$$P(\sigma^2; \sigma_0^2) = 1 + C_{n-1}(q_L \sigma_0^2 / \sigma^2) - C_{n-1}(q_U \sigma_0^2 / \sigma^2). \quad (4.15)$$

The three powers differ in how the quantiles q_L and q_U are computed.

Function pvalVAR contains two local functions: var.ql computes q_L and lnsAB solves equation (4.14). Figure 4.3 is similar to Figure 1.5 but instead of the CI we use the *p*-value to check the powers. Two versions of the *p*-value for the DL test are used: the α-dependent and the alternative computed by formulas (4.12) and (4.13) and depicted by a circle and a triangle, respectively. As follows from Theorem 4.6 both *p*-values produce the same power (the symbols coincide). The equal-tail *p*-value is computed by formula (4.11) and the unbiased *p*-value is

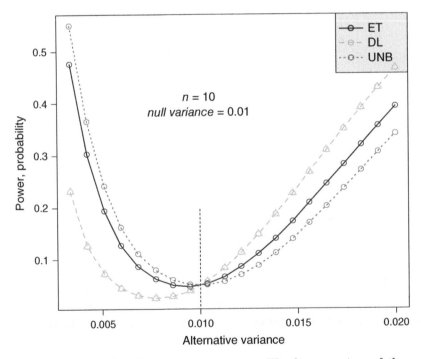

Figure 4.3: *Three tests for the normal variance. The line version of the power is computed by the formula (4.15) and points depict the simulation-derived power as the proportion of simulations for which the p-value is less than α. This graph was derived by the R function pvalVAR.*

computed by the same formula (4.12) but with q_L computed by the R function
var.ql. The simulation-derived power functions match the theoretical ones very
closely. As we saw previously, the DL test is biased and the ET and UNB tests
are close.

Example 4.10 ***Quality control involves constant testing of the normal***
variance of the product stability measurements. *The excepted variance of*
the measurements is $\sigma_0^2 = 0.1$ mm^2. To test that the variance remains the same
after the production system upgrade six measurements have been taken: 7.34,
6.95, 7.73, 7.23, 6.71, and 7.85 mm. Test the hypothesis that the variance did
not change by computing four p-values: the equal-tail test, the α-dependent and
the alternative for the DL test, and the unbiased test.

The computations are found in the function qual_cont. We use function
varTest of the package EnvStats for comparison.

```
> qual_cont()
                  p-value
EnvStats        0.17189484
Equal-tail      0.17189484
DL-alpha        0.09006665
DL-alternative  0.09210603
Unbiased        0.31998582
```

There is quite a difference between the p-values due to a small sample size,
$n = 6$. EnvStats and Equal-tail p-values coincide because they use the same
equal-tail test. The DL-alpha and DL-alternative p-values are close; however,
the p-value for the Unbiased approach is three times higher. Note that the
p-values are in the order depicted in Figure 4.2.

4.2 The overall powerful test

This section introduces an alternative to the DL and unbiased test: the overall
powerful test with the minimum area above the power function.

Definition 4.11 *We say that the test for the null hypothesis $H_0 : \theta = \theta_0$ versus*
$H_A : \theta \neq \theta_0$ is overall powerful (OP) if the area above the power function is the
minimum among all tests of the same size.

Note that the classic globally most powerful test as the test with maximum
power for each alternative is the OP test but the reverse is not true. It means
the OP test may serve as a second choice when the globally most powerful test
does not exist.

Our next step is to express the area of the test with the power given by (2.11) as the difference between the parameter means as formulated in the following lemma.

We need some general results similar to those from Demidenko (2020, p. 292).

Lemma 4.12 *(a) Let $F(x)$ and $f(x)$ be a cdf and pdf, and c be any number. Suppose that the mean exists, that is, $\left|\int_{-\infty}^{\infty} xf(x)dx\right| < \infty$. Then*

$$\int_{-\infty}^{c} F(x)dx = cF(c) - \int_{-\infty}^{c} xf(x)dx \tag{4.16}$$

$$\int_{c}^{\infty} (1 - F(x))dx = -c(1 - F(c)) + \int_{c}^{\infty} xf(x)dx. \tag{4.17}$$

(b) Let $F_X(x)$ and $F_Y(x)$ be the cdfs with pdfs $f_X(x)$ and $f_Y(x)$, respectively. Suppose that the means, μ_X and μ_Y, exist. Then

$$\int_{-\infty}^{\infty} (F_Y(x) - F_X(x))dx = \mu_X - \mu_Y. \tag{4.18}$$

Proof. (a) Apply integration by parts to the left-hand side of (4.16) by letting $u = F(x)$ and $dv = dx$. Then $du = f(x)dx$ and $v = x$. This implies what we intended to prove:

$$\int_{-\infty}^{c} F(x)dx = F(x)x|_{-\infty}^{c} - \int_{-\infty}^{c} xf(x)dx = cF(c) - \int_{-\infty}^{c} xf(x)dx.$$

Note that $\lim_{x \to -\infty} x F(x) = 0$ follows from the assumption that the mean exists. Now apply integration by parts to the left-hand side of (4.17) by letting $u = 1 - F(x)$:

$$\int_{c}^{\infty} (1 - F(x))dx = (1 - F(x))x|_{c}^{\infty} + \int_{c}^{\infty} xf(x)dx = -c(1 - F(c)) + \int_{c}^{\infty} xf(x)dx.$$

Again, $\lim_{x \to -\infty} x(1 - F(x)) = 0$ follows from the assumption that the mean exists.

(b) Express the left-hand side of (4.18) as

$$\int_{-\infty}^{\infty} (F_Y(x) - F_X(x))dx$$

$$= \int_{-\infty}^{c} (F_Y(x) - F_X(x))dx + \int_{c}^{\infty} (F_Y(x) - F_X(x))dx$$

$$= \int_{-\infty}^{c} (F_Y(x) - F_X(x))dx + \int_{c}^{\infty} [(1 - F_X(x)) - (1 - F_Y(x))]dx.$$

Apply (4.16) to the first term:

$$\int_{-\infty}^{c} (F_Y(x) - F_X(x))dx = cF_Y(c) - \int_{-\infty}^{c} xf_Y(x)dx - cF_X(c) + \int_{-\infty}^{c} xf_X(x)dx.$$

Now apply (4.17) to the second term:

$$\int_{c}^{\infty} [(1 - F_X(x)) - (1 - F_Y(x))]dx$$

$$= -c(1 - F_X(c)) + \int_{c}^{\infty} xf_X(x)dx + c(1 - F_Y(c)) - \int_{c}^{\infty} xf_Y(x)dx.$$

Add them up:

$$cF_Y(c) - \int_{-\infty}^{c} xf_Y(x)dx - cF_X(c) + \int_{-\infty}^{c} xf_X(x)dx - c(1 - F_X(c))$$

$$+ \int_{c}^{\infty} xf_X(x)dx + c(1 - F_Y(c)) - \int_{c}^{\infty} xf_Y(x)dx$$

$$= \mu_X - \mu_Y + cF_Y(c) + cF_X(c) - c - cF_X(c) + c - cF_Y(c)$$

$$= \mu_X - \mu_Y,$$

as was intended to prove.   □

Before formulating the next result we want to discuss the duality between the observational and parameter spaces connected through the cdf $F(s; \theta)$, where $s \in (-\infty, \infty)$ represents the observational and $\theta \in (-\infty, \infty)$ as the parameter space. Before, we viewed F as the cdf parametrized by θ. Now we switch the angle of interest by fixing the quantile, $s = q$, and considering F as a function of θ. We make the point that under assumptions 1–4 from Section 2.1, $1 - F(s; \theta)$ can be viewed as the cdf for θ with the pdf $-F'(q; \theta)$, where, as uniformly accepted in the book, the derivative means with respect to θ. Once, this point is clarified, one can view

$$\mu_q = - \int_{-\infty}^{\infty} \theta F'(q; \theta)d\theta$$

as the *parameter mean* at the given quantile q. Note that for a positive parameter, the integration domain is $(0, \infty)$. The following example illustrates the concept of the parameter mean.

Example 4.13 *Parameter mean for normal variance.*

For normal variance we have $F(q; \sigma^2) = C_{n-1}(q/\sigma^2)$. We want to see if $1 - F(q; \sigma^2) = 1 - C_{n-1}(q/\sigma^2)$ can be viewed as the cdf of σ^2 with the pdf $-F'(q; \sigma^2)$, where q hold fixed. Indeed, when $\sigma^2 \to 0$ then $1 - C_{n-1}(q/\sigma^2) \to 0$ and, when

$\sigma^2 \to \infty$, then $1 - C_{n-1}(q/\sigma^2) \to 1$, with the pdf $q/\sigma^4 c_{n-1}(q/\sigma^2)$. The parameter mean equals

$$\int_0^\infty \frac{q}{\sigma^2} c_{n-1}\left(\frac{q}{\sigma^2}\right) d\sigma^2 = \frac{q}{n-3}.$$

□

Now we are ready to formulate the main result.

Theorem 4.14 *Suppose that assumptions 1–4 listed in Section 2.1 hold. Then (a) the area above the power function defined by (2.11) can be expressed as*

$$\mu_U - \mu_L, \tag{4.19}$$

where μ_U and μ_L are referred to as the respective parameter means defined as

$$\mu_U = -\int_{-\infty}^\infty \theta F'(q_U; \theta) d\theta, \quad \mu_L = -\int_{-\infty}^\infty \theta F'(q_L; \theta) d\theta$$

under the assumption that the integrals are finite, and (b) the quantiles of the overall powerful test are found from the solution of the following system:

$$F(q_U; \theta_0) - F(q_L; \theta_0) = \lambda, \quad \frac{f(q_U; \theta_0)}{\int_{-\infty}^\infty f(q_U; \theta) d\theta} = \frac{f(q_L; \theta_0)}{\int_{-\infty}^\infty f(q_L; \theta) d\theta} \tag{4.20}$$

assuming that the integral $\int_{-\infty}^\infty f(q; \theta) d\theta$ exists for every q.

Proof. (a) The area above the power function (2.11) is

$$\int_{-\infty}^\infty (F(q_U; \theta) - F(q_L; \theta)) d\theta = \int_{-\infty}^\infty [(1 - F(q_L; \theta)) - (1 - F(q_U; \theta))] d\theta.$$

In connection with Lemma 4.12 we treat $1 - F(q_L; \theta)$ as F_Y and $1 - F(q_U; \theta)$ as F_X, and respectively $-F'(q_L; \theta)$ as f_Y and $-F'(q_U; \theta)$ as f_X, with θ treated as x. Now (4.19) follows from (4.18).

(b) We want to find q_L and q_U that minimize the area above the power function given by

$$\int_{-\infty}^\infty (F(q_U; \theta) - F(q_L; \theta)) d\theta$$

under the restriction

$$F(q_U; \theta_0) - F(q_L; \theta_0) = \lambda.$$

Introduce the Lagrange function

$$\mathcal{L}(q_L, q_U, \lambda) = \int_{-\infty}^\infty (F(q_U; \theta) - F(q_L; \theta)) d\theta - \nu(F(q_U; \theta_0) - F(q_L; \theta_0) - \lambda).$$

Differentiate with respect to q_L and q_U under the integral to obtain the necessary condition for the minimum

$$\int_{-\infty}^{\infty} f(q_U;\theta)d\theta = \nu f(q_U;\theta_0), \quad \int_{-\infty}^{\infty} f(q_L;\theta)d\theta = \nu f(q_L;\theta_0),$$

that leads to the system (4.20).   □

The OP test works the same as before: if F is the cdf of the test statistic S and $q_L < q_U$ are found from the solution of the system (4.20) the hypothesis $H_0 : \theta = \theta_0$ is rejected if the interval (q_L, q_U) does not cover S. Alternatively, in terms of cdfs, the hypothesis is accepted if

$$F(q_L;\theta_0) < F(S;\theta_0) < F(q_L;\theta_0).$$

The power of the OP test is given by formula (2.11).

This theorem is illustrated by our benchmark variance problem.

Example 4.15 *Normal variance and precision.*

We start with variance, σ^2. We aim to prove that the OP test for the normal variance is equivalent to the density level (DL) test from Section 2.3. In notation of Theorem 4.14, find

$$\int_0^{\infty} f(q;\theta)d\theta = \int_0^{\infty} \frac{1}{\sigma^2} c_{n-1}\left(\frac{q}{\sigma^2}\right) d\sigma^2.$$

A change of variable $u = q/\sigma^2$ gives

$$\int_0^{\infty} f(q;\theta)d\theta = \int_0^{\infty} \frac{1}{\sigma^2} c_{n-1}(u) \frac{\sigma^4}{q} du = \frac{1}{q} \int_0^{\infty} \sigma^2 c_{n-1}(u) du$$

$$= \int_0^{\infty} u^{-1} c_{n-1}(u) du = (n-3)^{-1}.$$

This means for normal variance the second equation in (4.20) turns into $f(q_U;\theta_0) = f(q_L;\theta_0)$. Consequently, the OP test is the same as the DL test because the quantiles are the same as follows from (2.8) and Example 2.14. We illustrate the optimality of the DL/OP test by plotting the area above the power function for three tests as a function of the null hypothesis variance, σ_0^2. Since the area is a monotonic function of σ_0^2 we plot the relative/normalized area as the area above the power divided by σ_0^2 for two sample sizes – see Figure 4.4 created by running `vartestOP()`. It is easy to prove that the relative area is constant – that is why the three curves turn into straight lines parallel to the x-axis. $n = 5$ represents a small sample size and $n = 20$ a moderate sample size. The DL test has the minimum area, and the unbiased test has the maximum area, and the area decreases with n (the power function is higher).

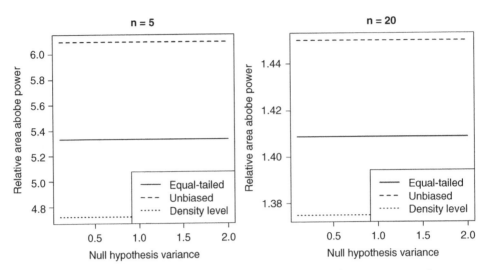

Figure 4.4: *The relative area above the power function for three tests for normal variance with two sample sizes (small and moderate). See Example 4.15. The density level and overall power tests are the same.*

Now we turn our attention to testing the precision parameter $p = 1/\sigma^2$ with the pdf defined as $f(q; p) = p c_{n-1}(pq)$. Then

$$\int_0^\infty f(q; \theta) d\theta = \int_0^\infty p c_{n-1}(pq) \, dp = (n-1)/q^2.$$

Note that unlike variance, the integral value depends on q, and therefore the OP test is different from the DL test. Specifically, the optimal quantiles for testing $H_0 : p = p_0$ are found by solving the system

$$C_{n-1}(p_0 q_U) - C_{n-1}(p_0 q_L) = \lambda, \quad q_L^2 c_{n-1}(p_0 q_L) - q_U^2 c_{n-1}(p_0 q_U) = 0.$$

This example teaches us that the OP test is not invariant with respect to reparametrization.   □

The following result generalizes the previous example.

Theorem 4.16 *(a) Let the location and scale pdfs be defined as*

$$f(s; \theta) = h(s - \theta) \ and \ f(s; \gamma) = \gamma^{-1} h(s\gamma^{-1}),$$

where h is a given pdf, $-\infty < \theta < \infty$ is the location, and γ is the positive scale parameter. Then the DL and OP tests coincide. (b) Let the pdf be defined as $f(s; \kappa) = \kappa h(\kappa s)$, where $\kappa > 0$ is the rate parameter. Then the OP quantiles are found by solving system

$$H(q_U \kappa_0) - H(q_L \kappa_0) = \lambda, \quad q_L^2 h(q_L \kappa_0) - q_U^2 h(q_U \kappa_0) = 0, \qquad (4.21)$$

where H is the respective cdf.

Proof. (a) For the location parameter, we have

$$\int_{-\infty}^{\infty} f(q;\theta)d\theta = \int_{-\infty}^{\infty} h(q-\theta)d\theta = \int_{-\infty}^{\infty} h(u)du = 1.$$

This implies that the OP and DL tests are equivalent and require finding quantiles by solving the system

$$H(q_U - \theta_0) - H(q_L - \theta_0) = \lambda, \quad h(q_U - \theta_0) = h(q_L - \theta_0).$$

For the scale parameter, after a change of variable, we have

$$\int_{-\infty}^{\infty} f(q;\theta)d\theta = \int_0^{\infty} \gamma^{-1} h(q\gamma^{-1})d\gamma = \int_0^{\infty} u^{-1} h(u)du,$$

which does not depend on q. Therefore the DL and OP tests coincide.

(b) For the rate parameter we have

$$\int_{-\infty}^{\infty} f(q;\theta)d\theta = \int_0^{\infty} \kappa h(\kappa q)d\kappa = q^{-2}\int_0^{\infty} uh(u)du = \mu/q^2,$$

where $\mu > 0$ is the mean. The system (4.21) follows directly from (4.20) using the previous formula. □

4.3 Duality: converting the CI into a hypothesis test

The principle of duality serves as the bridge between CI and hypothesis testing (Rao 1973, Lehmann and Romano 2005, Schervish 2012). According to this principle, a CI can be turned into a hypothesis test, and conversely a hypothesis test can be used to create a CI. In this section, we apply this principle to convert the CI to a dual test.

Given statistic S with the cdf $F(s;\theta)$ and pdf $f(s;\theta)$, the M-statistics finds the lower and the upper confidence limits $\theta_L = \theta_L(S)$ and $\theta_U = \theta_U(S)$ as solutions to the system of equations

$$F(S;\theta_L) - F(S;\theta_U) = \lambda, \; G(S;\theta_L) - G(S;\theta_U) = 0, \qquad (4.22)$$

where

$$G(s;\theta) = \begin{cases} \frac{\partial}{\partial\theta}F(s;\theta) = F'(s;\theta) \text{ for maximum concentration statistics} \\ \frac{\partial}{\partial s}F(s;\theta) = f(s;\theta) \text{ for mode statistics} \end{cases}.$$

Under conditions listed in the previous chapters, the CI (θ_L, θ_U) covers the true θ with probability λ: that is, $\Pr(\theta_L(S) < \theta < \theta_U(S)) = \lambda$ where $S \sim F(\cdot;\theta)$.

According to the duality principle we accept the null hypothesis $H_0 : \theta = \theta_0$ if the interval $(\theta_L(S), \theta_U(S))$ covers the true θ_0.

Theorem 4.17 *(a) If $(\theta_L(S), \theta_U(S))$ is the CI defined by (4.22), the dual test has the acceptance rule $\theta_L(S) \leq \theta_0 \leq \theta_U(S)$ with the power function given by*

$$P(\theta; \theta_0) = 1 + F(q_L; \theta) - F(q_U; \theta), \tag{4.23}$$

where the quantiles q_L and q_U are the solutions of the system

$$F(q_U; \theta_0) - F(q_L; \theta_0) = \lambda, \ G(q_U; \theta_0) - G(q_L; \theta_0) = 0. \tag{4.24}$$

Equivalently, the acceptance rule can be expressed as $q_L(\theta_0) \leq S \leq q_U(\theta_0)$. (b) The unbiased test and unbiased interval are dual, but the short CI and the DL test with the short acceptance interval are not.

Proof. (a) closely follows the proof of Theorem 2.3. The probability of acceptance is given by

$$\begin{aligned}
\Pr(\theta_L(S) \leq \theta_0 \leq \theta_U(S)) &= \Pr(\theta_L(S) \leq \theta_0) - \Pr(\theta_U(S) < \theta_0) \\
&= \Pr(S \leq \theta_L^{-1}(\theta_0)) - \Pr(S < \theta_U^{-1}(\theta_0)) \\
&= F(\theta_L^{-1}(\theta_0); \theta) - F(\theta_U^{-1}(\theta_0); \theta),
\end{aligned}$$

where θ_L and θ_U are understood as the function of S derived from (4.22), with $F(\cdot; \theta)$ as the cdf of S, and the superscript -1 indicates the function inverse. But due to the definition of the inverse from (4.22), we can rewrite

$$F(\theta_L^{-1}(\theta_0); \theta) - F(\theta_U^{-1}(\theta_0); \theta) = F(q_U; \theta_0) - F(q_L; \theta_0),$$

where θ_L and θ_U satisfy (4.24), so the probability complementary to the acceptance probability is the power given by (4.23). The interval $q_L(\theta_0) \leq S \leq q_U(\theta_0)$ is obtained through inverse of $\theta_L(S) \leq \theta_0 \leq \theta_U(S)$ using $q_L = \theta_U^{-1}(\theta_0)$ and $q_U = \theta_L^{-1}(\theta_0)$. The proof that the test dual to the unbiased CI is unbiased as well follows the proof of Theorem (3.6). The fact that the short CI and the DL test are not dual is demonstrated by the normal variance in the following example. □

This theorem proves the duality between the unbiased test and CI in MO-statistics. Although the unbiasedness of the CI and the respective test, in general, is well-known in statistics (Rao 1973, p. 470) Theorem 4.17 provides specific rules for building an unbiased test dual to the unbiased CI.

Theorem 4.17 applies to normal variance for the illustration as follows. See also Example 6.4 as an illustration of the multidimensional DL test.

Example 4.18 *Convert the short and unbiased CIs for normal variance into tests and check with simulations.*

The short CI for σ^2 is derived from (2.5) and defined through quantiles q_L and q_U found by solving (1.2) and (1.7). As discussed in Section 1.3 the solution is obtained by calling the function `var.ql` with the option `adj=-1`. The DL test for normal variance is derived in Example 2.14 and uses different quantiles computed using the option `adj=3`. Therefore the test dual to the short CI is different from the DL test which has a short acceptance interval. One can interpret this fact by saying that the minimization of the length of the CI in the parameter space is not equivalent to the minimization of the length of the acceptance interval in the observational space. See Figure 1.5 where the CIs and dual tests are compared via power functions tested by simulations. On the other hand, the unbiased CI and unbiased test use the same quantiles as discussed in Section 1.3. In other words, the unbiased CI and unbiased test are dual to each other.

4.4   Bypassing assumptions

The assumption that the test statistic and/or the parameter belong to the entire real line of Section 2.1 is easy to generalize to the finite or semi-infinite open interval domain. For example, when the parameter takes positive values, we treat 0 as $-\infty$, loosely speaking. In some statistical problems, function $F(s; \theta)$ is not a decreasing but an increasing function of θ. Then assumption 2 turns into $F'(s; \theta) > 0$. For example, F is an increasing function of the parameter in the exponential distribution parametrized by the rate parameter as $F(s; \lambda) = 1 - e^{-\lambda s}$. In this case, we still demand that the inflection point exists, and assumption 4 is replaced with $\lim_{\theta \to -\infty} F(s; \theta) = 0$ and $\lim_{\theta \to \infty} F(s; \theta) = 1$. The following example illustrates that statistical inference does not change much if $F' > 0$.

Example 4.19 *Statistical inference for the rate parameter in the exponential distribution.*

As in Example 2.9, we have a random sample of size n from an exponential distribution but now the parameter is the rate parameter $\lambda > 0$. The cdf of statistic S, the sum of observations, is given by $F(s; \lambda) = C_{2n}(2\lambda s)$. It is easy to see that $F'(s; \lambda) > 0$ and $\lim_{\lambda \to 0} F(s; \lambda) = 0$ and $\lim_{\lambda \to \infty} F(s; \lambda) = 1$. The condition (2.9) holds for the λ-parametrization:

$$\frac{d^2}{ds^2} \ln(2\lambda_0 c_{2n}(2\lambda_0 s)) = \frac{d^2}{ds^2}(0.5(2n-1)\ln s - \lambda_0 s)$$

$$= -\frac{2n-1}{2s^2} < 0.$$

The inequality (3.2) turns into

$$\frac{\partial^2}{\partial s \partial \lambda} \ln(2\lambda_0 c_{2n}(2\lambda_0 s))$$

$$= \frac{\partial^2}{\partial s \partial \lambda}(0.5(2n-1)\ln s - \lambda_0 s) = -1 < 0,$$

but the conclusion of Theorem 3.1 does not change. Consequently, the existence and uniqueness of MC and MO CIs and hypothesis testing equations hold.

An obvious derivation yields

$$\hat{\lambda}_{MC} = \frac{n - 0.5}{S}.$$

Note that the MC estimator of the rate parameter is not reciprocal to the MC estimator of the scale parameter. Development of the double-sided exact optimal CIs and hypothesis tests for λ are straightforward.   □

In some statistical problems, the first condition of assumption 4 does not hold, that is, the limit of $F(s; \theta)$ when $\theta \to -\infty$ ($\theta \to 0$ if the parameter is positive) exists but is less than 1. When this happens, for some observed S, the condition $F(S; \theta_L) - F(S; \theta_U) = \lambda$ may not be fulfilled, especially for a high-value λ, such as $\lambda = 0.95$, and therefore the exact CI does not exist. There is a simple remedy: replace the double-sided CI with the lower-sided CI $(-\infty, \theta_U^*)$, where θ_U^* is such that $F(S; \theta_U^*) = 1 - \lambda$. Note that since $\lim_{\theta \to \infty} F(s; \theta) = 0$ remains to be true θ_U^* exists. The following proves that such modification remains the coverage probability λ. Indeed, let $\lim_{\theta \to -\infty} F(s; \theta) = p(s)$ and $p_0 = \Pr(p(S) < \lambda)$ dependent on θ. Using | as conditional probability, the coverage probability of the modified CI is

$$\Pr(\theta < \theta_U^* \cap p(S) < \lambda) + \Pr(\theta_L < \theta < \theta_U \cap p(S) \geq \lambda)$$

$$= \Pr(\theta < \theta_U^* | p(S) < \lambda) \times p_0$$

$$+ \Pr(\theta_L < \theta < \theta_U | p(S) \geq \lambda) \times (1 - p_0) = \lambda p_0 + (1 - p_0)\lambda = \lambda,$$

as desired. We will be using this fix in several statistical problems, including the Poisson rate parameter and multiple correlation coefficient in the next chapter.

4.5   Overview

This book offers optimal exact statistical inference for double-sided CIs and hypothesis testing, which is especially valuable for asymmetric distributions with small samples. Two competing approaches to exact statistical inference are offered in this book under the umbrella of M-statistics: maximum concentration (MC) and mode (MO) statistics. Both approaches require statistic S having a distribution depending on an unknown single parameter θ. The two approaches use different criteria for the optimal CI and the double-sided hypothesis test. The MC-statistics produces the CI with minimum length (short CI) and the density level (DL) test with the minimum length (short) acceptance interval. The MO-statistics produces the unbiased CI and the dual unbiased test. In both paradigms, the estimator is derived as the limit point of the respective CI when the confidence level goes to zero. The MC estimate is the inflection point on the curve $F(S; \theta)$ as a function of θ given the observed S, where F is the cdf of S. From the probability perspective, the MC estimate is the point where the infinitesimal coverage probability of the unknown parameter, called the concentration probability, is maximum. The MO estimate maximizes the density, and it turns into the maximum likelihood estimate when S is the sufficient statistic. In this case, the MO estimator maximizes the cumulative information function. Summary equations are presented in Table 4.1.

M-statistics is especially valuable for asymmetric distributions with a small sample size. For example, the length of the short CI for variance is 25% smaller than the standard equal-tail CI when $n = 6$.

Under mild conditions, the equations for the optimal CI limits and hypothesis test quantiles exist and are unique. The numerical solutions are effectively found by Newton's algorithm starting from the equal-tail confidence limits and test quantiles. When the derivatives required for Newton's algorithm are difficult to compute, general optimization algorithms, such as those implemented in the R function `optimize` or solution of a nonlinear system of equations from the package `nleqslv` can be employed.

Classic statistics constructs the CI on the additive scale when the quality of the interval estimation is measured by its length. For positive parameters, such as variance and standard deviation, the relative scale, such as the log scale, is more appropriate by keeping the lower limit of the CI away from zero.

The p-value is well-defined for a symmetric distribution or a one-sided test. The definition of the p-value for the double-sided test is challenging and not unique. We offer a definition that complies with the two conditions: (a) the distribution of the p-value is uniform under the null hypothesis, and (b) the probability that $p < \alpha$ equals the power of the test under the alternative. For the unbiased test, p depends on α, but for the equal-tail and DL test an alternative a-independent definition of the p-value is proposed.

According to the dual principle, an exact CI can be converted to an exact hypothesis test. The unbiased CI leads the unbiased test, but the short CI does not produce the DL test with the minimum length acceptance interval.

To ensure the exact coverage probability the limit of the cdf when the parameter approaches the left boundary of the parameter domain must be 1 for any observed value of the test statistic. When the limit is smaller than 1, there is a simple fix: replace the double-sided with the lower-sided CI.

Table 4.1　Summary of the optimal double-sided exact statistical inference.

	Short CI	Density level (DL) test	MC estimator, $\lambda \to 0$
MC statistics	$F(S; \theta_L) - F(S; \theta_U) = \lambda$ $F'(S; \theta_L) - F'(S; \theta_U) = 0$	$F(q_U; \theta_0) - F(q_L; \theta_0) = \lambda$ $f(q_L; \theta_0) - f(q_U; \theta_0) = 0$	$F''(S; \theta) = 0$
	Unbaised CI	**Unbaised test**	**MO estimator, $\lambda \to 0$**
MO statistics	$F(S; \theta_L) - F(S; \theta_U) = \lambda$ $f(S; \theta_L) - f(S; \theta_U) = 0$	$F(q_U; \theta_0) - F(q_L; \theta_0) = \lambda$ $F'(q_L; \theta_0) - F'(q_U; \theta_0) = 0$	$f'(S; \theta) = 0$

p-value

$$p = \begin{cases} \dfrac{\alpha}{F(q_L; \theta_0)} F(S; \theta_0) & \text{if } F(S; \theta_0) \le \dfrac{1}{\alpha} F(q_L; \theta_0) \\[2ex] \dfrac{\alpha}{\alpha - F(q_L; \theta_0)} (1 - F(S; \theta_0)) & \text{if } F(S; \theta_0) > \dfrac{1}{\alpha} F(q_L; \theta_0) \end{cases}$$

Power function

$$P(\theta; \theta_0) = 1 + F(q_L; \theta) - F(q_U; \theta)$$

$F(.;\theta)$ and $f(.;\theta)$ are the cdf and pdf of statistic S with a single parameter θ

θ_L and θ_U are the lower and upper confidence limits dependent on S

q_L and q_U are the lower and upper quantiles dependent on θ_0

The double-sided hypothesis is $H_0 : \theta = \theta_0$ versus the alternative $H_A : \theta \ne \theta_0$

$\lambda = 1 - \alpha$ is the confidence and α is the significance level of the double-sided test

$'$ is the derivative with respect to parameter θ

The MC estimate is the inflection point on the curve $F(S; \theta)$ as a function of θ

The MO estimate maximizes the pdf over θ given the observed S

References

[1] Demidenko, E. (2020). *Advanced Statistics with Applications in R*. Hoboken, NJ: Wiley.

[2] Dunne, A., Pawitan, Y., and Doody, L. (1996). Two-sided p-values from discrete asymmetric distributions based on uniformly most powerful unbiased tests. *The Statistician* 45: 397–405.

[3] Fisher, R.A. (1925). *Statistical Methods for Research Workers*. Edinburgh: Oliver and Boyd.

[4] Johnson, N.L., Kotz, S., and Balakrishnan, N. (1994). *Continuous Univariate Distributions*, vol. 1, 2e. New York: Wiley.

[5] Lehmann, E.L. and Romano, J.P. (2005). *Testing Statistical Hypothesis*, 3e. New York: Springer.

[6] Rao, C.R. (1973). *Linear Statistical Inference and Its Applications*, 2e. New York: Wiley.

[7] Schervish, M. (2012). *Theory of Statistics*. New York: Springer.

[8] Wasserstein, R.L. and Lazar, N.A. (2016). The ASA's statement on p-values: Context, process, and purpose. *American Statistician* 70: 129–133.

Chapter 5

M-statistics for major statistical parameters

This chapter displays M-statistics in action: the maximum concentration (MC) and mode (MO) exact statistical inference are applied to major statistical parameters. Here M-statistics operates with CIs, tests, and point estimators under one methodological umbrella. We develop new exact optimal double-sided CIs and statistical tests and provide numerical algorithms to find the solutions with R codes found on GitHub. When the closed-form for the cdf does not exist we express the cdf via integral over the density and use numerical integration (integrate) to facilitate derivation with respect to the parameter. Real-life examples illustrate the application and how to call the respective R functions. An important aspect of our discussion is the derivation of the power function of the test, which is valuable from a theoretical perspective to compare the tests, and from an application point of view to the sample size determination given the power value. We do not provide a comprehensive literature review on statistical inference of parameters considered in this chapter but illustrate M-statistics.

5.1 Exact statistical inference for standard deviation

As was mentioned in the motivating example, the optimal CIs for normal variance were developed many years ago. Remarkably, the exact statistical inference for normal standard deviation is still underdeveloped. For example, the widespread CI for σ, as the square root of the equal-tail CI for variance, does not have minimum length and is not unbiased. Certainly, when the sample size is large

M-statistics: Optimal Statistical Inference for a Small Sample, First Edition. Eugene Demidenko.
© 2023 John Wiley & Sons, Inc. Published 2023 by John Wiley & Sons, Inc.

the difference between the equal- and unequal-tail approaches is negligible but when n is small the difference may be up to 20%, as illustrated shortly.

The goal of this section is to derive optimal statistical inference for normal standard deviation using the general results of the previous chapters. Although the inference is very similar to the normal variance we repeat the major steps due to the importance of this parameter and to make our coverage self-contained. As in the case of normal variance, statistic $S = \sum (Y_i - \overline{Y})^2$ is used for our development. Remember that the cdf and pdf of statistic S are $F(s; \sigma) = C_{n-1}(s\sigma^{-2})$ and $f(s; \sigma) = \sigma^{-2} c_{n-1}(s\sigma^{-2})$, respectively, but now parametrized by σ, not σ^2, in contrast to normal variance.

5.1.1   MC-statistics

As mentioned previously, the lower and upper limits of the traditional equal-tail $100\lambda\%$ CI for σ are $\sqrt{S/q_{1-\alpha/2}}$ and $\sqrt{S/q_{\alpha/2}}$, where $q_{1-\alpha/2}$ and $q_{\alpha/2}$ are the $(1 - \alpha/2)$ and $\alpha/2$ quantiles of the chi-square distribution with $n - 1$ df. As follows from Section 2.2 and Example 2.4, the short double-sided CI (σ_L, σ_U) is derived by solving the following system of equations:

$$C_{n-1}(S\sigma_L^{-2}) - C_{n-1}(S\sigma_U^{-2}) - \lambda = 0,$$
$$\sigma_L^{-3} c_{n-1}(S\sigma_L^{-2}) - \sigma_U^{-3} c_{n-1}(S\sigma_U^{-2}) = 0. \tag{5.1}$$

Upon notation $S\sigma_L^{-2} = q_U$ and $S\sigma_U^{-2} = q_L$, after simple algebra, we arrive at an equivalent but a simplified system:

$$C_{n-1}(q_U) - C_{n-1}(q_L) - \lambda = 0, \ n \ln(q_U/q_L) - (q_U - q_L) = 0. \tag{5.2}$$

After q_L and q_U are found, the exact short CI for σ takes the familiar form $(\sqrt{S/q_U}, \sqrt{S/q_L})$. Note that this CI does not account for the positiveness of σ as in Example 7 of Section 2.5.1 (the log-scale CI is discussed below). Similarly to the variance, the optimal quantiles q_L and q_U do not depend on the observed S. Newton's algorithm solves the system (5.2) starting from the equal-tail chi-square quantiles implemented in the previously discussed R function var.ql with adj=0. It requires only three or four iterations to converge.

Example 5.1 *Five normally distributed observations are* $2.3, 4.1, 3.5, 5.2,$ $1.4.$ *Compute the equal-tail and short 95% CI for* σ.

The call of the sd.ciex function and its output are shown here.

```
> sd.ciex()
              Equal-tail CI    Short CI
Lower limit       0.893693  0.7328118
Upper limit       4.286320  3.5868385
% CI sigma short = 19
```

Note that the short CI reduces the length by moving the limits downward. The limits of the traditional equal-tail CI are $\sqrt{S/q_{0.975}} = 0.8937$ and $\sqrt{S/q_{0.025}} = 4.2863$. The equal-tail CI is 19% wider than the short CI. ☐

Now we turn our attention to testing $H_0 : \sigma = \sigma_0$ versus $H_A : \sigma \neq \sigma_0$. The power function of the test is given by

$$P(\sigma; \sigma_0) = 1 + C_{n-1}(q_L \sigma_0^2/\sigma^2) - C_{n-1}(q_U \sigma_0^2/\sigma^2),$$

where the quantiles, q_L and q_U are σ_0-independent. The null hypothesis is rejected if the interval $(\sqrt{S/q_U}, \sqrt{S/q_U})$ does not cover σ_0.

The short CI has quantiles computed by solving the system (5.2). The density level (DL) test requires the solution of the system (2.8)

$$C_{n-1}(\sigma_U \sigma_0^{-2}) - C_{n-1}(\sigma_L \sigma_0^{-2}) - \lambda = 0,$$

$$c_{n-1}(\sigma_L \sigma_0^{-2}) - c_{n-1}(\sigma_U \sigma_0^{-2}) = 0. \tag{5.3}$$

Again, upon notation $\sigma_L \sigma_0^{-2} = q_L$ and $\sigma_U \sigma_0^{-2} = q_U$, we arrive at a simplified system

$$C_{n-1}(q_U) - C_{n-1}(q_L) - \lambda = 0, \ (n-3)\ln(q_U/q_L) - (q_U - q_L) = 0. \tag{5.4}$$

This system is numerically solved by the same R function `var.ql` using `adj=3`. The DL test rejects the null hypothesis if σ_0 is outside the interval $(\sqrt{S/q_U}, \sqrt{S/q_L})$, which can be treated as the CI dual to the DL test. The power function of this test along with the equal-tail test discussed previously for $\sigma_0 = 1$ and $n = 10$ is depicted in Figure 5.1. The unbiased test is discussed shortly. This figure, similar to Figure 1.5, was created with the R function `sdM`. The power values are checked by simulations (number of simulations = 1 million) using the respective CI and rejection rule. The symbols at $\sigma = 0.85$ depict the proportion of simulations for which the dual CIs do not cover the true value σ_0, and the symbols at $\sigma = 1.1$ depict the proportion of simulations for which the null hypothesis is rejected according to the rejection rule. The computation of the p-value is discussed in Section 5.1.4.

All four tests are exact but just for the unbiased test the tangent line is horizontal to the x-axis at $\sigma_0 = 1$ (the unbiased test is discussed in Section 5.1.3).

The MC estimator for σ is found from equation $F''(S; \sigma) = 0$, which is equivalent to the maximization of

$$\ln(-F'(S; \sigma)) = \text{const} + \frac{n-3}{2} \ln \frac{S}{\sigma^2} - \frac{S}{2\sigma^2} - 3\ln\sigma.$$

After some algebra, we find $\hat{\sigma}_{MC} = \sqrt{S/n}$, the same as the maximum likelihood estimator. It is instructive to prove that the short CI defined by (5.1) converges to $\hat{\sigma}_{MC}$. When $\lambda \to 0$ from the first equation of (5.2), we deduce that $q_L \to q_0$ and

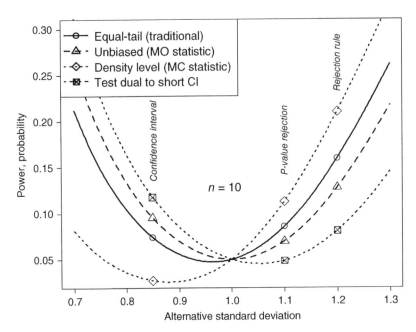

Figure 5.1: *The power functions of equal- and unequal-tail tests for the two-sided standard deviation hypothesis $H_0 : \sigma_0 = 1$ with $n = 10$ and $\alpha = 0.05$. The symbols depict the power values derived through simulations (number of simulations = 1 million). This graph was generated by the R function* sdM.

$q_U \to q_0$, where q_0 is a subject of determination. Dividing the second equation in (5.2) by $q_U - q_L$, due to the Mean Value Theorem we obtain

$$1 = n \lim_{q_L \to q_0, q_U \to q_0} \frac{\ln q_U - \ln q_L}{q_U - q_L}$$

$$= n \lim_{q_L \to q_0, q_U \to q_0} \frac{1/q_*(q_U - q_L)}{q_U - q_L} = \frac{n}{q_0}$$

where $q_L < q_* < q_U$. Thus $q_0 = n$ and therefore when $\lambda \to 0$, the lower and upper limits converge to $\sqrt{S/q_0} = \sqrt{S/n}$.

5.1.2   MC-statistics on the log scale

As we argued in Section 2.5.1, the length of the CI must be measured on the relative scale for a positive parameter such as standard deviation. As a special case of (5.2), the second equation of (5.2) turns into

$$(n+1)\ln(q_U/q_L) - (q_U - q_L) = 0 \tag{5.5}$$

which defines the short CI for σ on the log scale. Again, this system can be solved b y calling the var.ql function with adj=-1. When $\lambda \to 0$ the CI on the log scale shrinks to $\widehat{\sigma}_{MCL} = \sqrt{S/(n+1)}$, with proof similar to the short CI.

5.1.3 MO-statistics

The quantiles of the unbiased test, as follows from equations (3.4), are found by solving the system

$$C_{n-1}(q_U) - C_{n-1}(q_L) - \lambda = 0, \ (n-1)\ln(q_U/q_L) - (q_U - q_L) = 0,$$

which is numerically solved by `var.ql` with `adj=1`. The null hypothesis is rejected if the unbiased CI $(\sqrt{S/q_U}, \sqrt{S/q_L})$ does not cover σ_0. When $\lambda \to 0$, the CI shrinks to the MO estimator, $\widehat{\sigma}_{MO} = \sqrt{S/(n-1)}$.

5.1.4 Computation of the p-value

The computation of the p-value for testing σ is very similar to computation for σ^2, as presented in Section 4.1.2, computed by the R function `pvalVAR`. The p-value for standard deviation is checked through simulations in Figure 5.1. The symbol at $\sigma = 1.1$ depicts the proportion of simulations for which the p-value is less than $\alpha = 0.05$. The equal-tail p-value is computed by formula (4.6). For other tests formula (4.3) applies. The theoretical power function and the power based on the p-values derived from simulations are in very good agreement.

5.2 Pareto distribution

The Pareto distribution has been known for more than 150 years and has a wide field of applications including sociology, seismology, and ecology (Arnold 2015). One of the most important problems is the statistical inference of the parameters. Surprisingly, there are no exact statistical inferences for the rate parameter of the Pareto distribution to date (Johnson et al. 1994). The primary goal of this section is to fill the gap and employ M-statistics to derive the exact optimal confidence interval and hypothesis testing for the simplest and yet fundamental one-parameter Pareto distribution. The secondary goal of this analysis is to demonstrate the optimal properties of MO- and MC-statistics based on their criteria.

The pdf and cdf of the Pareto distribution take the form

$$f(x; a) = \frac{a}{x^{a+1}}, \quad F(x; a) = 1 - \frac{1}{x^a}, \quad 1 \le x < \infty, \quad a > 0,$$

where the matter of our concern is the rate parameter a. Note that for small a, the mean and variance may not exist. The joint pdf of the random sample $X_1, ..., X_n$ from this distribution takes the form

$$f(x_1, ..., x_n; a) = \frac{a^n}{(\Pi x_i)^{a+1}} = a^n e^{-(a+1)\sum \ln x_i},$$

where

$$S = \sum_{i=1}^{n} \ln X_i$$

is the sufficient statistic (we will refer to S as the *sumlog* statistic). Malik (1966) proved that the sumlog statistic follows a gamma distribution with the shape parameter n and the rate parameter a: that is, the pdf of S is given by

$$g(s; a) = \frac{a^n}{\Gamma(n)} s^{n-1} e^{-as}, \quad s > 0.$$

The proof follows from two facts: (a) if X has the pdf a/x^{a+1} then $\ln X$ has the gamma distribution with the shape parameter 1 and the rate parameter $a + 1$, and (b) the sum of n gamma-distributed independent random variables with the same rate parameter is gamma-distributed with the shape parameter n and the same rate parameter. This result is instrumental for our exact optimal small-sample statistical inference for the Pareto parameter a because the pdf and cdf of the gamma distributions (dgamma and pgamma) are built-in functions in R that substantially facilitate our algorithms. Note that the cdf of S is an increasing function of a, but it does not affect our optimal inference.

5.2.1   Confidence intervals

Here we derive the unbiased and short CIs for the Pareto parameter a using sumlog statistic S. We start with the equal-tail CI because it is used as a starting point in Newton's algorithm.

The limits of the equal-tail CI for a are found by solving the equation

$$G_n(S; a) = \eta,$$

where G_n denotes the cdf of the gamma distribution with the shape parameter n and the rate parameter a. The value $\eta = (1 - \lambda)/2$ is used for the lower and $\eta = (1 + \lambda)/2$ for the upper exact confidence limit. Newton's algorithm takes the generic form

$$a_{k+1} = a_k - \frac{G_n(S; a_k) - \eta}{G_n'(S; a_k)}, \quad k = 0, 1, \dots$$

Express the derivative of the gamma distribution with respect to the rate parameter a as the difference between the cdfs of two gamma distributions as follows:

$$\begin{aligned}
G_n'(S; a) &= n \int_0^S \frac{a^{n-1}}{\Gamma(n)} s^{n-1} e^{-as} ds - \int_0^S \frac{a^{n-1}}{\Gamma(n)} s^n e^{-as} ds \\
&= \frac{n}{a} \int_0^S \frac{a^n}{\Gamma(n)} s^{n-1} e^{-as} ds - \frac{\Gamma(n+1)}{a\Gamma(n)} \int_0^S \frac{a^{n+1}}{\Gamma(n+1)} s^n e^{-as} ds \\
&= \frac{n}{a} (G_n(S; a) - G_{n+1}(S; a)).
\end{aligned} \tag{5.6}$$

Finally iterations take the form

$$a_{k+1} = a_k - \frac{(G_n(S; a_k) - \eta)a_k}{n(G_n(S; a_k) - G_{n+1}(S; a_k))} \quad k = 0, 1, \ldots$$

We found that starting from the ML estimate, $a_0 = n/S$, is successful for both lower and upper confidence limits: see our R function `pareto`. The densities and cdfs are plotted by calling `pareto(job=0)` and checked through simulations. The call `pareto(job=1)` illustrates the convergence of Newton's algorithm graphically, and the call `pareto(job=1.1)` checks the iterations via the coverage probability.

The unbiased CI for a solves the system

$$G_n(S; a_U) - G_n(S; a_L) = \lambda, \quad n \ln a_U/a_L - (a_U - a_L)S = 0. \tag{5.7}$$

Note that we simplified the density equation by taking the log transformation. Using the previously derived formula (5.6), we obtain the adjustment vector for (a_L, a_U):

$$\begin{bmatrix} -\frac{n}{a_L}(G_n(S; a_L) - G_{n+1}(S; a_L)) & \frac{n}{a_U}(G_n(S; a_U) - G_{n+1}(S; a_U)) \\ -n/a_L + S & n/a_U - S \end{bmatrix}^{-1}$$

$$\times \begin{bmatrix} G_n(S; a_U) - G_n(S; a_L) - \lambda \\ n \ln a_U/a_L - (a_U - a_L)S \end{bmatrix}.$$

The call `pareto(job=2)` implements this algorithm for a simulated random sample of size n, and `pareto(job=2.1)` checks the coverage probability. Note that we use the vectorized version of the algorithm and for this purpose, the solution of the system (5.7) is obtained in the element-wise form.

Since S is the sufficient statistic, as follows from Theorem 3.10, the MO estimator of a coincides with the ML estimator:

$$\widehat{a}_{MO} = \widehat{a}_{ML} = \frac{n}{S}.$$

This result sheds new light on the small-sample property of the ML estimator as the limit point of the unbiased CI.

Now we turn our attention to the maximum concentration (short) CI that solves the system

$$G_n(S; a_U) - G_n(S; a_L) = \lambda, \quad G_n'(S; a_U) - G_n'(S; a_L) = 0.$$

Newton's algorithm requires $G_n''(S; a)$. The repeat of formula (5.6) gives the desired result:

$$G_n''(S; a) = \frac{n}{a}(G_n(S; a) - G_{n+1}(S; a))' = \frac{n}{a}\left[\frac{n}{a}(G_n(S; a) - G_{n+1}(S; a))\right]$$

$$-\frac{n}{a}\left[\frac{n+1}{a}(G_{n+1}(S;a) - G_{n+2}(S;a))\right]$$

$$= \frac{n}{a^2}[nG_n(S;a) - (2n+1)G_{n+1}(S;a) + (n+1)G_{n+2}(S;a)]. \quad (5.8)$$

Newton's algorithm has a familiar form:

$$\begin{bmatrix} -G_n'(S;a_L) & G_n'(S;a_U) \\ -G_n''(S;a_L) & G_n''(S;a_U) \end{bmatrix}^{-1} \begin{bmatrix} G_n(S;a_U) - G_n(S;a_L) - \lambda \\ G_n'(S;a_U) - G_n'(S;a_L) \end{bmatrix},$$

where the first and second derivatives of G_n are computed by formulas (5.6) and (5.8), respectively.

Figures 5.2 and 5.3 compare the three CIs visually derived by calling `pareto(job=4)` and `pareto(job=5)`, respectively. The former plots the length of CIs as a function of the sample size with true $a = 2$ (number of simulations = 100,000). Not surprisingly, the short CI has a minimal length, however, the equal-tail CI is shorter than the unbiased CI. The second figure compares the relative length as the length of the CI divided by the true a plotted on the x-axis for $n = 10$ (the default sample size). Interestingly, the relative length is almost constant, but the order of CIs in terms of the length remains the same.

Now we discuss the computation of the MC estimate of a as the solution of the equation $G_n''(S;a) = 0$. As follows from (5.8) $\widehat{a}_{MC}$ is the solution of

$$nG_n(S;a) - (2n+1)G_{n+1}(S;a) + (n+1)G_{n+2}(S;a) = 0.$$

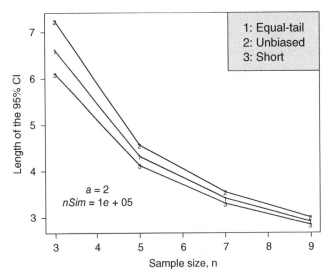

Figure 5.2: *The length of three 95% CIs for a as a function of the sample size, n. This graph was created by calling* `pareto(job=4)`.

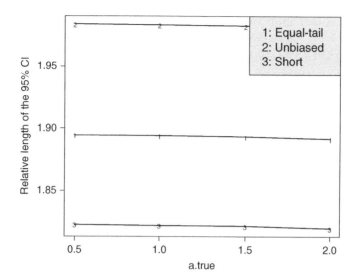

Figure 5.3: *Comparison of the relative length of three 95% CIs as a function of the true Pareto parameter, a, for $n = 10$. This graph was created by calling* `pareto(job=5)`.

Since we have a good starting point $\widehat{a}_{MO} = n/S$ Newton's algorithm applies again:

$$a_{k+1} = a_k - \frac{nG_n - (2n+1)G_{n+1} + (n+1)G_{n+2}}{G_n'''},$$

where $(S; a_k)$ is omitted for brevity. After omitting some tedious algebra we obtain

$$G_n''' = a^{-1}(n^2 G_n - 3n(n+1)G_{n+1} + (3n^2 + 6n + 3)G_{n+2} - (n^2 + 3n + 2)G_{n+3}).$$

Remarkably, $\widehat{a}_{MC} \simeq \widehat{a}_{MO}$ for all combinations of a and n: see `pareto(job=6)`.

5.2.2 Hypothesis testing

We aim to test the double-sided null hypothesis $H_0 : a = a_0$ versus $H_A : a \neq a_0$. As in the preceding section, three tests are discussed. The simplest equal-tail test computes the quantiles q_L and q_U using `qgamma` function as `qgamma((1-lambda)/2,shape=n, rate=a0)` and `qgamma((1+lambda)/2, shape=n, rate=a0)`, respectively. The null hypothesis is rejected if the observed S falls outside of (q_L, q_U) with the power function

$$P(a; a_0) = 1 - G_n(q_U; a) + G_n(q_L; a). \tag{5.9}$$

The formula for the power function remains the same but with different quantiles depending on the test.

The unbiased test finds quantiles by solving the system

$$G_n(q_U; a_0) - G_n(q_U; a_0) = \lambda, \quad G'_n(q_U; a_0) - G'_n(q_U; a_0) = 0.$$

The adjustment vector for (q_L, q_U) in Newton's algorithm takes the familiar form

$$\begin{bmatrix} -G'_n(q_L; a_0) & G'_n(q_U; a_0) \\ -G''_n(q_L; a_0) & G''_n(q_U; a_0) \end{bmatrix}^{-1} \begin{bmatrix} G_n(q_U; a_0) - G_n(q_L; a_0) - \lambda \\ G'_n(q_U; a_0) - G'_n(q_L; a_0) \end{bmatrix},$$

where G'_n and G'''_n are computed by formulas (5.6) and (5.8) starting from the equal-tail quantiles.

The density level test solves the system

$$G_n(q_U; a_0) - G_n(q_U; a_0) = \lambda, \quad (n-1)\ln(q_U/q_L) - a_0(q_U - q_L) = 0$$

using Newton's adjustment

$$\begin{bmatrix} -G'_n(q_L; a_0) & G'_n(q_U; a_0) \\ -(n-1)/q_L + a_0 & (n-1)/q_U - a_0 \end{bmatrix}^{-1} \begin{bmatrix} G_n(q_U; a_0) - G_n(q_L; a_0) - \lambda \\ (n-1)\ln(q_U/q_L) - a_0(q_U - q_L) \end{bmatrix}.$$

Figure 5.4 was created by running `pareto(job=7)`. It is similar to Figure 1.5 for normal variance from Section 1.3 and depicts three power functions with the

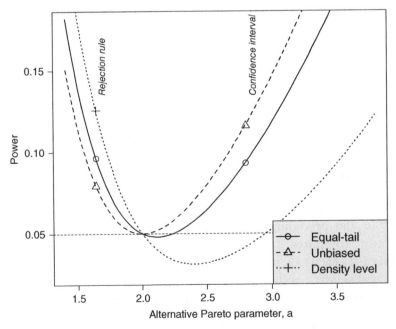

Figure 5.4: *Power functions of three exact tests ($n = 10$, $a_0 = 2$). Symbols represent values from simulations using the respective rejection rule and dual confidence interval.*

symbol values to the left of $a_0 = 2$ derived from simulations using the rejection rule and the dual CIs to the right of a_0 ($n = 10$ and $\lambda = 0.95$). All three tests are exact – the power curves pass the point $(2, 0.05)$. The equal-tail and DL tests are biased – the tangent line at $a = a_0$ has a negative slope, but the tangent line of the unbiased power curve is horizontal. The DL test has a heavy bias. The power value is the proportion of simulations for which $S < q_L$ or $S > q_U$ and the dual CI (the null hypothesis is rejected if the CI does not cover the true a_0). Note that we intentionally plotted the power functions in the proximity of the null value to emphasize the bias.

5.3 Coefficient of variation for lognormal distribution

The coefficient of variation (CV), or the relative standard deviation, hereafter denoted as ζ, is defined as the ratio of the standard deviation to the mean. This coefficient naturally emerges in the lognormal distribution: we say that random variable X has a lognormal distribution if $\ln X \sim \mathcal{N}(\mu, \sigma^2)$. It is well known that the CV of the lognormal distribution does not depend on μ and is expressed through σ^2 as follows:

$$\zeta = \sqrt{e^{\sigma^2} - 1}. \tag{5.10}$$

Note that although CV frequently applies to normally distributed data, as we discuss later in Section 5.6, it has aberrant properties when the mean is close to zero, while the CV defined in (5.10) is properly defined. We emphasize its connection to σ because due to the well-known approximation $e^x - 1 \simeq x$ for small x it is easy to see that $\zeta \simeq \sigma$. This approximation helps interpret σ as the relative standard deviation of the original data (Demidenko 2020, p. 118). For example, $\sigma = 0.1$ means approximately the standard deviation of the lognormal distribution is 10% of the mean.

Several authors developed approximate methods for CIs and hypothesis testing for ζ based on maximum likelihood, including Cohen (1951), but no exact statistical inference for ζ exists up to date.

The goal of this section is to apply MO-statistics to develop the exact optimal statistical inference for ζ using the statistic $S = \sum_{i=1}^{n} (\ln X_i - \overline{\ln X})^2$. We enjoy the invariance property of the MO inference since the CV for the lognormal distribution is a monotonic transformation of the normal variance, and as such, the results of Section 1.3 are readily applicable.

Expressing σ^2 through the parameter of interest as $\sigma^2 = \ln(1 + \zeta^2)$, the cdf of S is written as

$$F(s; \zeta) = C_{n-1}(s/\ln(1 + \zeta^2)), \; s > 0.$$

It is elementary to check that the cdf complies with all four assumptions from Section 2.1 for positive ζ. Since ζ is a strictly increasing reparametrization of the normal variance we apply the MO-statistics by the respective transformation of the unbiased CI and statistical test. Specifically, if $\sigma_L^2 = S/q_U$ and $\sigma_U^2 = S/q_L$, where q_L and q_U are the solutions of system (1.11), the $100\lambda\%$ limits of the unbiased CI for ζ are $\sqrt{e^{\sigma_L^2} - 1}$ and $\sqrt{e^{\sigma_U^2} - 1}$.

The exact unbiased testing of the null hypothesis $H_0 : \zeta = \zeta_0$ versus $H_A : \zeta \neq \zeta_0$ is equivalent to testing the normal variance with the null hypothesis $H_0 : \sigma^2 = \sigma_0^2 = \ln(1 + \zeta_0^2)$. The hypothesis is rejected if σ_0^2 is outside of the interval $(S/q_U, S/q_L)$ with the power function defined by

$$P(\zeta; \zeta_0) = 1 + C_{n-1}(q_L \ln(1 + \zeta_0^2)/\ln(1 + \zeta^2))$$
$$- C_{n-1}(q_U \ln(1 + \zeta_0^2)/\ln(1 + \zeta^2)). \qquad (5.11)$$

Again, thanks to the invariance property, the MO estimator for ζ is computed as

$$\widehat{\zeta}_{MO} = \sqrt{e^{\widehat{\sigma}_{MO}^2} - 1} = \sqrt{e^{\widehat{\sigma}^2} - 1}, \qquad (5.12)$$

where $\widehat{\sigma}^2 = S/(n-1)$ is the traditional unbiased estimator of the normal variance using the log observations.

Here we compare the MO statistical inference with the established asymptotic inference that uses the same estimator (5.12) along with its asymptotic variance

$$var(\widehat{\zeta}_{MO}) \simeq \frac{\sigma^4 e^{2\sigma^2}}{2(n-1)(e^{\sigma^2} - 1)}$$

as discussed by Niwitpong (2013) and Hasan and Krishnamoorthy (2017). This formula can be easily derived from the delta method using the fact that $var(\widehat{\sigma}^2) = 2\sigma^4/(n-1)$.

Figure 5.5 depicts the two power functions for $n = 10$, $\zeta_0 = 0.3$ and $\alpha = 0.05$. This figure is derived by the R function cvLN. The exact power of the unbiased test is computed by (5.11) and verified by simulations using the unbiased CI depicted by a circle, that is, the power is the proportion of simulations for which ζ_0 is outside of the unbiased interval $(S/q_U, S/q_L)$. Values q_L and q_U are computed by the function var.ql with adj=1. The asymptotic power values, depicted by a triangle, are derived as the proportion of simulations for which the true CV is outside of the asymptotic interval $\widehat{\zeta}_{MO} \pm \Phi^{-1}(1 - \alpha/2)\sqrt{var(\widehat{\zeta}_{MO})}$, with σ^2 replaced with its unbiased estimate, $S/(n-1)$. The simulation-derived asymptotic power is neither unbiased nor exact. There is a substantial loss of power to the right of the null value.

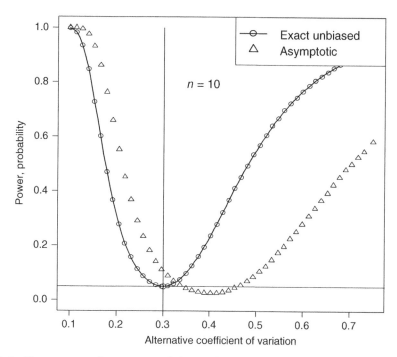

Figure 5.5: *Two power functions of the MO exact unbiased and asymptotic tests for the two-sided test of the lognormal CV. Circles depict the dual power values derived from the respected simulated CIs and triangles depict the power from the asymptotic CIs.*

5.4 Statistical testing for two variances

Comparison of variances from two independent normal samples has almost a 100-year-long history, starting from Wilks (1932). For example, in engineering and technology, the respective test is a part of the quality control system – the centerpiece of Six Sigma introduced by Motorola in 1986. The exact test relies on the F-distribution as the quotient of the unbiased variance estimates from the two samples. Other tests, including Bartlett's test, are available – we refer to Keating and Tripathi (1984), Lim and Loh (1996), and Zhou et al. (2020) for further reading. In this section, we will work with the F-test. It is common to use the equal-tail approach for testing that the two variances are the same – for example, this approach is realized in the R built-in function var.test. As we mentioned before, although this test is exact, it does not possess optimal properties such as being unbiased. The development of an unbiased test is the goal of this section.

We want to test the equality of the variances from two independent normally distributed samples of different sizes $X_i \overset{\text{iid}}{\sim} \mathcal{N}(\mu_1, \sigma_1^2)$ and $Y_j \overset{\text{iid}}{\sim} \mathcal{N}(\mu_2, \sigma_2^2)$, where $i = 1, 2, ..., n_1$ and $j = 1, 2, ..., n_2$. The null hypothesis is two-sided: $H_0 : \sigma_1^2 = \sigma_2^2$ against $H_A : \sigma_1^2 \neq \sigma_2^2$. If $\widehat{\sigma}_1^2$ and $\widehat{\sigma}_2^2$ denote the unbiased estimators of the two variances (sample variances), the quotient of the variances under the null hypothesis follows F-distribution with $n_1 - 1$ and $n_2 - 1$ dfs. If α is the significance level (typically, $\alpha = 0.05$) the equal-tail F-test computes the lower and upper quantiles, $q_{\alpha/2}$ and $q_{1-\alpha/2}$, of the F-distribution. Then, if the quotient of the variances falls outside of the interval $(q_{\alpha/2}, q_{1-\alpha/2})$, the null hypothesis is rejected.

We argue that the tail probabilities must not necessarily be the same. Moreover, they can be found to make the test unbiased as outlined in Section 3.1. To make the derivation concise it is customary to equivalently express the null hypothesis as $H_0 : \nu = 1$ versus $H_A : \nu \neq 1$, where the test statistic is the same as previously, $S = \widehat{\sigma}_1^2 / \widehat{\sigma}_2^2$, and $\nu = \sigma_1^2 / \sigma_2^2$. This reformulation allows us to write the cdf of the test statistic as

$$\Pr(\widehat{\sigma}_1^2 / \widehat{\sigma}_2^2 \leq x) = \mathcal{F}(x/\nu), \; x > 0,$$

where $\mathcal{F}$ denotes the cdf of the F-distribution with dfs $n_1 - 1$ and $n_2 - 1$. The condition on the unbiasedness of the test based on the F-distribution was derived by Ramachandran (1958). Next we formulate this condition in a simplified form readily available for a numerical solution following Demidenko (2020, p. 570).

Theorem 5.2 *Properties of the F-test. Let $q_1 < q_2$ be the quantiles of the F-distribution with dfs $n_1 - 1$ and $n_2 - 1$ such that $\mathcal{F}(q_2) - \mathcal{F}(q_1) = 1 - \alpha$. Then the power function of the F-test for the null hypothesis $\nu = 1$ versus $\nu \neq 1$ is given by*

$$P(\nu) = 1 - \mathcal{F}(q_2/\nu) + \mathcal{F}(q_1/\nu),$$

where $\nu = \sigma_1^2 / \sigma_2^2$ is the alternative variance quotient. The test with the equal-tail probabilities, where $q_1 = \mathcal{F}^{-1}(\alpha/2)$ and $q_2 = \mathcal{F}^{-1}(1 - \alpha/2)$, is unbiased when $n_1 = n_2$. To make the F-test unbiased in the general case, the quantiles q_1 and q_2 must be found by solving the following couple of equations:

$$\mathcal{F}(q_2) - \mathcal{F}(q_1) - (1 - \alpha) = 0,$$

$$k_1 \ln \frac{q_2}{q_1} - (k_1 + k_2) \ln \frac{k_2 + k_1 q_2}{k_2 + k_1 q_1} = 0, \qquad (5.13)$$

where $k_1 = (n_1 - 1)/2$ and $k_2 = (n_2 - 1)/2$.

The difference between the equal-tail and unbiased F-test is negligible for large sample sizes – see Example 5.3. Newton's algorithm for solving the system of equations (5.13) is presented in Demidenko (2020) and implemented in the R function `q1q2F`. This function returns a two-dimensional array with components q_1 and q_2 as the solution of the system (5.13), respectively. Figure 5.6 displays the power functions of the unbiased and equal-tail F-tests for $n_1 = 3$ and $n_2 = 10$

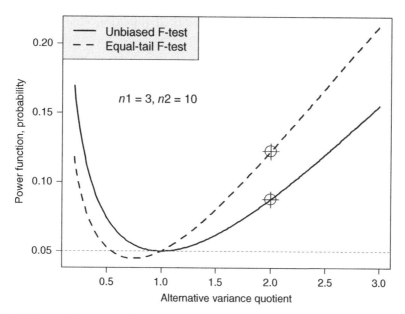

Figure 5.6: *Power functions for unbiased and equal-tail F-tests with small sample sizes. Circles indicate simulation-derived power values based on the rejection rule and crosses depict the power values based on the p-value. This graph was created by calling* `smallF(job=1)`.

with $\alpha = 0.05$. Both tests are exact and the curves pass the point $(1, 0.05)$, but the slope of the power function of the equal-tail test is positive at $\nu = 1$, and therefore the power to the left of 1 is smaller than 0.05. Of note, we intentionally use small sample sizes and a tight neighborhood around $\nu_0 = 1$ to underline the biasedness of the equal-tail test.

The theoretical powers are checked through simulations (number of simulations = 1 million) using the rejection rule (circles) and p-value (crosses). The match is very close (details of the p-value computation are presented shortly).

5.4.1 Computation of the p-value

As follows from Section 4.1, the p-value for the equal-tail test according to (4.6) is computed as

$$p = 2 \times \min(\mathcal{F}(S), 1 - \mathcal{F}(S)).$$

The p-value for the unbiased test is computed as

$$p = \min\left(\frac{\alpha\mathcal{F}(S)}{\mathcal{F}(q_L)}, \frac{\alpha(1 - \mathcal{F}(S))}{\alpha - \mathcal{F}(q_L)}\right).$$

The two power functions are tested via simulations using these p-values at alternatives $\nu = 0.5$ and $\nu = 2$ in Figure 5.6. The crosses depict the proportion of simulations for which $p < \alpha$. The theoretical rejection rule and the p-value derived power values match each other.

5.4.2   Optimal sample size

The power function enables computing the optimal sample size for achieving the desired power given the alternative variance quotient ν. In the following example, we are seeking a low-cost design with a minimum total sample size $n_1 + n_2$ that achieves 80% and 90% power given the alternative. The example illustrates the computations.

Example 5.3 *(a) A manufacturing company tests the precision of two technological processes by comparing variances in two batches of sample sizes $n_1 = 40$ and $n_2 = 81$ with variance estimates $\widehat{\sigma}_1^2 = 5.2$ and $\widehat{\sigma}_2^2 = 4.1$ on the scale of microns squared. Test the hypothesis that the true variances are the same using the p-value with $\alpha = 0.05$ for equal-tail and unbiased F-tests. (b) Compute the optimal/low-cost batch sample sizes to achieve the power 80% and 90% for the unbiased test with the alternative variance quotient $\nu = 2$ and the size of the test $\alpha = 5\%$.*

(a) The call `smallF(job=2)` returns p-values pET $= 0.3684$, pUNB $= 0.3979$. The p-values are fairly close but the sample sizes are too small to claim statistical significance using the traditional cut-off value of 0.05. (b) Now we want to know the sample size for each batch to achieve the desired power of detection of the alternative variance quotient $\nu = 2$. To find optimal n_1 and n_2, we must solve the optimization problem $\min(n_1 + n_2)$ under restriction $P(2) = p_0$, where n_1 and n_2 are treated as continuous and $p_0 = 0.8$ or $p_0 = 0.9$. We solve this problem numerically by computing power P as a matrix on the grid of values over the two sample sizes. Then we find n_1 and n_2 for each contour corresponding to $P(2) = p_0$ such that $n_1 + n_2$ takes the minimal value. As it easy to prove with the Lagrange multiplier technique, the optimal sample sizes are where the tangent line to the contour has slope -1. The graph with the two powers as functions of n_1 and n_2 is presented in Figure 5.7 and created by calling `smallF(job=3)`. The sample sizes must be increased to ensure that the probability of rejection of the alternative (beta-error) is small. The fact that the optimal batches have similar sizes is explained by the fact that the sample sizes are large which makes the power close to symmetric with respect to n_1 and n_2.

5.5   Inference for two-sample exponential distribution

We aim to develop optimal statistical inference for the comparison of the rate parameters using two independent samples from the exponential distribution

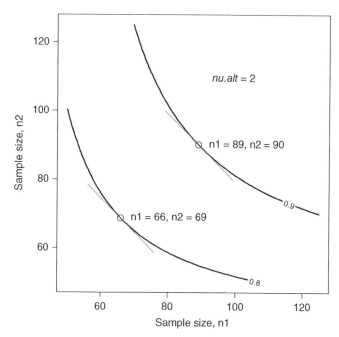

Figure 5.7: *The optimal sample size yields the desired power of detection with the alternative quotient $\nu = 2$. This graph was created by calling the function* `smallF(job=3)`*; see Example 5.3.*

parametrized as $\lambda e^{-\lambda x}$, where λ is referred to as the rate parameter. This inference is important in many engineering applications including quality control (Levine et al. 2001).

The unbiased statistical test for exponential distribution is very similar to that developed in the previous section for variances, but now the null value is not necessarily equal 1. Besides testing, we develop short and unbiased CIs.

We have two independent samples $\{X_i, i = 1, ..., n_1\}$ and $\{Y_j, j = 1, ..., n_2\}$ from exponential distributions with the rate parameters λ_1 and λ_2, respectively. We aim to develop an optimal double-sided test for $H_0 : \lambda_2/\lambda_1 = \nu_0$ versus $\lambda_2/\lambda_1 \neq \nu_0$ and construct the optimal CI for $\nu = \lambda_2/\lambda_1$, where ν_0 may be different from 1, unlike in the previous section, however, in many applications we are interested in testing whether the two rates are the same, $\lambda_1 = \lambda_2$, that is, $\nu_0 = 1$. Note that in the traditional hypothesis testing setting, we express the alternative hypothesis on the additive scale with the null hypothesis written as $\lambda_2 - \lambda_2 = \delta$. However, for statistical problems with positive parameters, including the two-sample exponential problem, it is advantageous to use the relative scale, that is, write the null hypothesis as $\lambda_2/\lambda_1 = \nu_0$.

Although the topic of statistical inference for the two-sample exponential distribution has a long history and is well covered (Krishnamoorthy 2006) we concentrate on optimal tests and confidence intervals that are especially valuable for unbalanced design when $n_1 \neq n_2$ with small n.

The basis for the exact statistical inference for $\nu = \lambda_2/\lambda_1$ is the fact that

$$\frac{\overline{X}}{\nu \overline{Y}} \sim F(2n_1, 2n_2), \tag{5.14}$$

the F-distribution with df $2n_1$ and $2n_2$ with the density

$$f(x; 2n_1, 2n_2) = cx^{n_1-1}\left(1 + \frac{n_1}{n_2}x\right)^{-(n_1+n_2)}, \tag{5.15}$$

where $c > 0$ is the normalizing coefficient. The left-hand side can be viewed as a pivotal quantity. Equivalently, as we do in M-statistics, we work with statistic $S = \overline{X}/\overline{Y}$ having cdf $\mathcal{F}(\nu^{-1}s; 2n_1, 2n_2)$. Although statistical testing for ν is similar to the ratio of two normal variances we discuss the short confidence interval and the MC estimator of ν as well.

5.5.1 Unbiased statistical test

Let $q_1 = q_{\alpha/2}$ and $q_2 = q_{1-\alpha/2}$ be the quantiles of the F-distribution with $2n_1$ and $2n_2$ df. The null hypothesis is accepted if $\nu_0^{-1}S$ falls within the interval $(1/q_2, 1/q_1)$. The equal-tail test is exact but biased. The following theorem presents the power of the test along with the definition of unequal-tail quantiles that yield an unbiased test.

Theorem 5.4 *Let $q_1 = q_{\alpha/2}$ and $q_2 = q_{1-\alpha/2}$ be the quantiles of the F-distribution with $2n_1$ and $2n_2$ df. The equal-tail F-test rejects the null hypothesis $H_0 : \nu = \nu_0$ versus $H_0 : \nu \neq \nu_0$ if S falls outside the interval $(\nu_0/q_2, \nu_0/q_1)$ with the power function given by*

$$P(\nu; \nu_0) = 1 - \mathcal{F}(q_2\nu_0/\nu; 2n_1, 2n_2) + \mathcal{F}(q_1\nu_0/\nu; 2n_1, 2n_2). \tag{5.16}$$

This test is unbiased if and only if the sample sizes are the same, $n_1 = n_2$. If the sample sizes are not the same, the quantiles must be found from the system of equations

$$\mathcal{F}(q_2; 2n_1, 2n_2) - \mathcal{F}(q_1; 2n_1, 2n_2) - (1 - \alpha) = 0,$$

$$n_1 \ln(q_2/q_1) - (n_1 + n_2)\ln(n_2 + n_1q_2)/(n_2 + n_1q_1) = 0. \tag{5.17}$$

to make the test unbiased with the power function (5.16).

Proof. Since in the notation of Section 3.1, $F(s; \theta) = \mathcal{F}(\nu^{-1}s)$ we have $F'(s; \theta) = -\nu_0^{-2}sf(\nu_0^{-1}s)$ the unbiased test defined by equations (3.1) requires the solution of the system

$$\mathcal{F}(\nu_L^{-1}\nu_0) - \mathcal{F}(\nu_U^{-1}\nu_0) - (1-\alpha) = 0,$$
$$\nu_0^{-2}\nu_L f(\nu_0^{-1}\nu_L) - \nu_0^{-2}\nu_U f(\nu_0^{-1}\nu_U) = 0,$$

where $f = \mathcal{F}'$ and the dfs $2n_1, 2n_2$ are omitted for simplicity. Multiply the second equation by ν_0, and denote $q_2 = \nu_L^{-1}\nu_0$ and $q_2 = \nu_U^{-1}\nu_0$. Then the previous system takes the form

$$\mathcal{F}(q_2) - \mathcal{F}(q_1) - (1-\alpha) = 0, \quad q_1 f(q_1) - q_2 f(q_2) = 0.$$

The second equation simplifies upon using the expression for the density (5.15) and taking the log transform – this yields (5.17). The adjustment vector in Newton's algorithm takes the form

$$\begin{bmatrix} -f(q_1) & f(q_2) \\ -\frac{n_1}{q_1} + \frac{n_1(n_1+n_2)}{n_2+n_1 q_1} & \frac{n_1}{q_2} - \frac{n_1(n_1+n_2)}{n_2+n_1 q_2} \end{bmatrix}^{-1} \begin{bmatrix} b_1 \\ b_2 \end{bmatrix},$$

where $f(q_1)$ and $f(q_2)$ are the respective pdf values, and b_1 and b_2 are the left-hand sides of equations (5.17), with the equal-tail quantiles as starting values. Typically, it takes three-four iterations to converge (see the function q12). Note that the optimal quantiles do not depend on ν_0.

An example of a power function for two tests, equal- and -unequal (unbiased), is depicted in Figure 5.8 with $n_1 = 1$, $n_2 = 10$, and $\nu_0 = 1$. See the function exp2 with the function q12 as part of it. The functions are shown in the proximity of the null value $\nu_0 = 1$ for better visualization of the unbiasedness of the test. Symbols represent the simulation-derived powers based on the rejection rule.

The power function (5.16) can be used to compute the optimal sample sizes in each group that minimize $n_1 + n_2$ to achieve the desired power. Figure 5.9 illustrates these computations for the null $\nu_0 = 1$ and the alternative $\nu = 2$. As follows from this graph, to achieve power 80% with the size $\alpha = 5\%$, one needs $n_1 = 32$ observations from sample X and $n_2 = 34$ observations from sample Y. The total sample size $n_1 + n_2 = 66$ is the minimum among all combinations of n_1 and n_2 that yield a power of 80%. Note that if n_1 and n_2 are treated as continuous variables the optimal point is where the $-45°$ line touches the contour of the power function. To achieve 90% power, one needs 44 observations per group. □

Example 5.5 *Insider trading of stock by an individual who receives a piece of confidential information about the company is criminal.* *John Smith sold his stock of $1 million one minute after the CEO of a pharmaceutical company announced that its leading anticancer drug causes severe side effects. Smith was accused of inside trading. In the past, there were six occasions when*

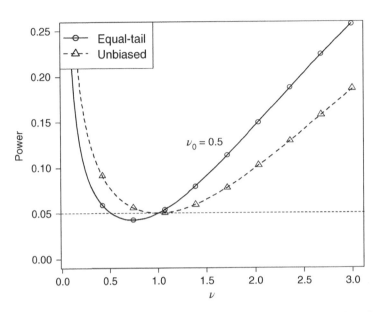

Figure 5.8: *The power functions for the two tests: equal-tail and unbiased. The equal-tail test is exact but biased – the power function does not touch the horizontal line $\alpha = 0.05$. This graph was created with the function* exp2(job=1).

this stock was sold by individuals after a major announcement of company set-back: 6, 3, 5, 2, 10, and 4 minutes. Assuming that the trading time after a major public announcement follows an exponential distribution test the hypothesis that Smith's trading time rate is the same as that of the other six traders (a) by applying the rejection rule and (b) by computing the p-value.

(a) In the language of the previous discussion we have $n_1 = 1$ and $n_2 = 6$ with the null hypothesis $H_0 : \nu = \nu_0 = 1$. The observed statistic is

$$S = \frac{\overline{X}}{\overline{Y}} = \frac{1}{(6 + 3 + 5 + 2 + 10 + 4)/6} = 0.2.$$

The R function q12 solves the system (5.17). The call q12(n1=1,n2=6) returns q1=0.03896439 and q2=6.55555052 (the R function q12 is a local function of exp2). (a) The interval $(1/q2, 1/q1) = (0.152, 25.7)$ contains the observed value $S = 0.2$. This means we cannot reject the null hypothesis that John Smith's trading time rate is the same as the other six traders. (b) The p-value for the unbiased test is computed using formula (4.3) with the target significance level $\alpha = 0.05$.

5.5.2   Confidence intervals

The goal of this section is to develop short and unbiased confidence intervals for ν. We start with the exact equal-tail CI: due to (5.14), it takes the form $(q_2^{-1}S, q_1^{-1}S)$,

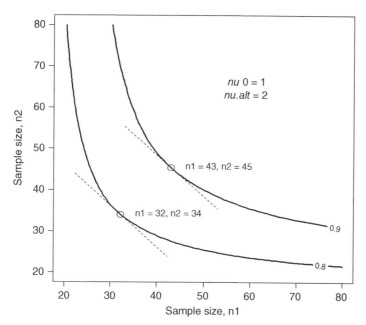

Figure 5.9: *Optimal n_1 and n_2 to achieve the powers 0.8 and 0.9. The line $n_1 + n_2 = const$ is tangent to the power curve at the optimal solutions.*

where q_1 and q_2 are $\alpha/2$ and $(1 - \alpha/2)$ quantiles of the F-distribution with $2n_1$ and $2n_2$ dfs. Since the derivative of the cdf of S with respect to ν takes the form $-s\nu^{-2}f(\nu^{-1}s)$, the short $100\lambda\%$ CI (ν_L, ν_U) is derived from the solution of the following system of equations:

$$\mathcal{F}(S\nu_L^{-1}) - \mathcal{F}(S\nu_U^{-1}) - \lambda = 0, \ S\nu_L^{-2}f(S\nu_L^{-1}) - S\nu_U^{-2}f(S\nu_U^{-1}) = 0,$$

where S is the observed statistic. After multiplying the second equation by S and letting $q_U = \nu_L^{-1}S$ and $q_L = \nu_U^{-1}S$, we arrive at a slightly simplified system to solve:

$$\mathcal{F}(q_U) - \mathcal{F}(q_L) - \lambda = 0, \ q_U^2 f(q_U) - q_L^2 f(q_L) = 0.$$

Note that q_L and q_U do not depend on statistic S. This system can be further simplified by using the expression for the density of the F-distribution:

$$\mathcal{F}(q_U; 2n_1, 2n_2) - \mathcal{F}(q_L; 2n_1, 2n_2) - \lambda = 0,$$

$$(n_1 + d) \ln(q_L/q_U) - (n_1 + n_2) \ln \frac{n_2 + n_1 q_L}{n_2 + n_1 q_U} = 0, \qquad (5.18)$$

where $d = 1$ is the adjustment parameter. Once q_L and q_U are found we compute the short CI as $(q_U^{-1}S, q_L^{-1}S)$. Again, we use Newton's algorithm starting from $q_L = q_{\alpha/2}$ and $q_U = q_{1-\alpha/2}$. We notice that the system (5.18) is the same as (5.17)

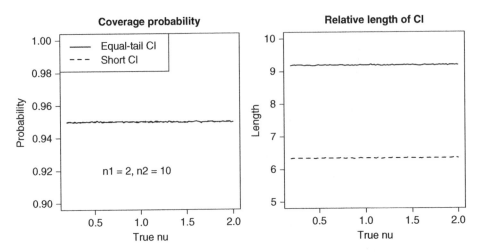

Figure 5.10: *Simulation derived coverage probability and relative length of the equal-tail and short-length CIs for $n_1 = 2$ and $n_2 = 10$. This graph was created by calling* `exp2(job=3)`.

except for -1 in the first coefficient of the second equation. Therefore, it is convenient to use d as the argument of Newton's algorithm in the R function. Then for the system (5.17), we have $d = 0$, and for the system (5.18), we have $d = 1$.

For a positive parameter such as the ratio of two rates, ν, it is advantageous to measure the length of the interval on the log scale – see Section 2.5.1.

The equal-tail and short-length CIs are compared visually in Figure 5.10 for $n_1 = 2$ and $n_2 = 10$ with $\lambda = 0.95$ using simulations. The coverage probability is very close to nominal. The length of the interval is computed on the relative scale with respect to the true ν as $(S/q_U - S/q_L)/\nu$. Several observations can be made: (1) the MC interval is considerably shorter on the entire range of true ν, (2) the relative length is close to constant, and (3) the relative length is quite large due to a small number of observations.

5.5.3 The MC estimator of ν

Since estimators of the rates are reciprocal of the means, the maximum likelihood estimator of ν is

$$\widehat{\nu}_{ML} = \frac{\overline{Y}^{-1}}{\overline{X}^{-1}} = \frac{\overline{X}}{\overline{Y}} = S.$$

The MC estimator is the limit point of the MC CI when the confidence level approaches zero. As follows from Theorem 2.6 the MC estimator is the maximizer of the function $\nu^{-2} f(\nu^{-1} S)$. To shorten the notation we let $x = \nu^{-1} S$ so that the

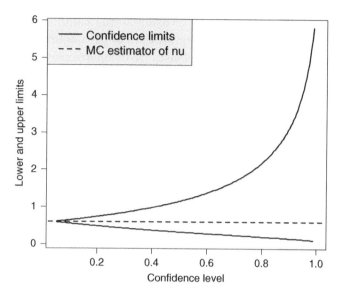

Figure 5.11: *Illustration of the MC estimator of ν with $n_1 = 2$ and $n_2 = 10$ when $\lambda \to 0$ for the short CI with $S = 1$.*

function to maximize is $x^2 f(x)$. To simplify maximization, we take the log and omit constants – this leads to maximization

$$(n_1 + d)\ln x - (n_1 + n_2)\ln(n_2 + xn_1), \quad x > 0,$$

where $d = 1$. To find the maximizer take the derivative and set it to zero. Simple algebra yields the MC estimator

$$\widehat{\nu}_{MC} = \frac{n_1(n_2 - 1)}{n_2(n_1 + 1)} S.$$

Remember that the MC estimator has a maximum density around the true ν and is computed as the inflection point of the cdf as a function of the parameter evaluated at the observed S – see Theorem 2.6. It is easy to see that $\widehat{\nu}_{MC} < \widehat{\nu}_{ML}$, and when n_1 and n_2 are large, the ML and MC estimators are close.

Figure 5.11 illustrates the MC estimator $\widehat{\nu}_{MC}$ as the limit point of the short CI when $\lambda \to 0$ with $n_1 = 2$, $n_2 = 10$, and $S = 1$.

5.6 Effect size and coefficient of variation

The goal of this section is to illustrate the M-statistics by developing the exact statistical inference for the effect size (ES) and its reciprocal, the coefficient of variation (CV). As in the previous section, we enjoy the invariance property of the

MO-statistics to effortlessly translate the exact statistical inference developed for the ES to the CV. ES and CV are major statistical parameters, but small sample inference is underdeveloped, to say the least. We develop exact short and unbiased confidence intervals and respective double-sided statistical tests for any sample size n. Although the two parameters are reciprocally related, the ES is easier to handle because it has a noncentral t-distribution while the CV is a discontinuous function when the mean equals zero. Thus, our strategy is to develop the theory for the ES first and then apply it to the CV using the reciprocal property although not without a caveat because of the discontinuity. Working out effective numerical algorithms by solving nonlinear integral equations is an important part of our development.

5.6.1 Effect size

Effect size, sometimes called the standardized mean, is a popular metric in statistical applications defined as

$$\theta = \frac{\mu}{\sigma},$$

where μ and σ are the mean and standard deviation of the normally distributed general population (Cohen 1988). ES emerges as a major parameter in power functions of many hypothesis tests regarding the means of the normal distribution such as one- or two-sample t-tests under the equal variance assumption. ES is a popular metric in medical applications as the treatment effect is measured on the scale of standard deviation. Wolfe and Hogg (1971) advocated reporting the ES on the probability scale. This measure was rediscovered under different names: common language ES (McGraw and Wong 1992) and D-value (Demidenko 2016, 2020).

Let $\{Y_i, i = 1, 2, ..., n\}$ be a random sample from the normal distribution $\mathcal{N}(\mu, \sigma^2)$. Traditionally, the ES is estimated as $\widehat{\theta} = \overline{Y}/\widehat{\sigma}$, referred to as the sample ES, where $\widehat{\sigma} = \sqrt{\sum(Y_i - \overline{Y})^2/(n-1)}$. This estimator is positively biased. Indeed, due to the independence of $\overline{Y}$ and $\widehat{\sigma}^2$, we have $E(\widehat{\theta}) = E(\overline{Y})E(\widehat{\sigma}^{-1}) = \mu E(\widehat{\sigma}^{-1})$, but

$$E(\sigma\widehat{\sigma}^{-1}) = \sqrt{n-1} \int_0^\infty s^{-1/2} c_{n-1}(s)ds = \frac{\sqrt{n-1}\,\Gamma(n/2-1)}{\sqrt{2}\,\Gamma(n/2-1/2)},$$

and therefore

$$E(\widehat{\theta}) = \theta \frac{\sqrt{n-1}\,\Gamma(n/2-1)}{\sqrt{2}\,\Gamma(n/2-1/2)}.$$

One can prove that $E(\widehat{\theta}) > \theta$ by employing Gautschi's inequality (Gautschi 1958):

$$\frac{\Gamma(n/2-1)}{\Gamma(n/2-1/2)} > \frac{\sqrt{2}}{\sqrt{n-2}}.$$

Finally, the bias of the sample ES follows from

$$E(\widehat{\theta}) > \theta\sqrt{(n-1)/(n-2)} > \theta.$$

For example, for $n = 5$, the bias is about 25%. Using the previous expression for $E(\widehat{\theta})$, it is obvious how to adjust $\widehat{\theta}$ to make the estimator unbiased. Regardless of this attempt, the distribution of $\widehat{\theta}$ remains asymmetric, and therefore M-statistics is relevant.

Previous authors (Zou 2007, Browne 2010) developed an equal-tail CI for θ – the software is available in the R package MBESS (this package will be used shortly). The goal of this section is to apply M-statistics to derive the exact short and unbiased CIs, the respective two-sided statistical tests, and MC and MO estimators for θ. Our derivation uses a well-known fact that statistic

$$S = \sqrt{n}\widehat{\theta}$$

has a noncentral t-distribution with $n - 1$ df and the noncentrality parameter $\sqrt{n}\theta$ (Lehmann and Romano 2005). Although the cdf of the noncentral distribution is a built-in function in many statistical packages including R, we prefer the integral representation because our theory requires derivation with respect to the noncentrality parameter, which is easier to obtain by differentiation of the pdf under the integral.

Short CI

Denote $\mathcal{T}$ and $\mathcal{T}'$ the cdf and its derivative with respect to the noncentrality parameter of the noncentral t-distribution with $n - 1$ df, respectively (remember that throughout this book, sign $'$ indicates differentiation with respect to the parameter, unless noted otherwise), where $F(s; \theta) = \mathcal{T}(s; \text{df} = n - 1, \text{ncp} = \sqrt{n}\theta)$ is computed by the R function pt. The MC (short) CI (2.2) for θ is derived by solving the system of equations

$$\mathcal{T}_L - \mathcal{T}_U - \lambda = 0, \quad \mathcal{T}'_L - \mathcal{T}'_U = 0, \qquad (5.19)$$

where to shorten the notation, the arguments are omitted and the subscripts L and U indicate that $\mathcal{T}$ and its derivative are evaluated at $(S; \sqrt{n}\theta_L)$ and $(S; \sqrt{n}\theta_U)$, respectively. Since we need the first and second derivatives of this function with respect to the noncentrality parameter an integral representation of the cdf of the noncentral t-distribution is beneficial. Specifically, rewrite $S = V/\sqrt{U}$, where $V \sim \mathcal{N}(\sqrt{n(n-1)}\theta, n-1)$ and $U \sim \chi^2(n-1)$, are independent. Then the cdf of S is $\Pr(S \le s) = \Pr(V - s\sqrt{U} < 0)$. Using the convolution formula, we obtain an integral representation of the cdf of the noncentral t-distribution:

$$F(s; \theta) = \mathcal{T}(s; \theta) = 2\int_0^\infty w\Phi(u(s))c_{n-1}(w^2)dw, \qquad (5.20)$$

where

$$u(s) = sw/\sqrt{n-1} - \sqrt{n}\theta,$$

and Φ and c_{n-1} are the standard normal cdf and chi-square pdf, respectively. This integral, as well as other integrals hereafter, are numerically evaluated by the R function `integrate`. The first two derivatives with respect to θ are easily obtained by differentiation under the integral:

$$\mathcal{T}' = -2\sqrt{n} \int_0^\infty w\phi(u(s))c_{n-1}(w^2)dw,$$

$$\mathcal{T}'' = -2n \times \int_0^\infty w(u(s))\phi(u(s))c_{n-1}(w^2)dw. \tag{5.21}$$

The equal-tail upper limit θ_U is computed by solving the equation $\mathcal{T}_U(S; \sqrt{n}\theta_U) = \alpha/2$ by Newton's algorithm with the adjustment

$$\frac{\mathcal{T}(S; \sqrt{n}\theta_U) - \alpha/2}{\mathcal{T}'(S; \sqrt{n}\theta_U)},$$

starting from the normal approximation suggested by Johnson and Welch (1940). The lower limit is computed similarly. Alternatively, the parallel-chord algorithm can be applied when the derivative is replaced with its lower bound (Ortega and Rheinboldt 2000, p. 181). To find the lower bound for $\mathcal{T}'$, we notice that $\phi \leq 1/\sqrt{2\pi}$, and therefore

$$\mathcal{T}' \geq -\frac{2\sqrt{n}}{\sqrt{2\pi}} \int_0^\infty wc_{n-1}(w^2)dw = -\frac{\sqrt{n}}{\sqrt{2\pi}}$$

because the integral (after the change of variable $v = w^2$) equals 0.5. The parallel-chord algorithm uses the adjustment $-\sqrt{2\pi}(\mathcal{T}(S; \sqrt{n}\theta) - \alpha/2)/\sqrt{n}$. This method may take more iterations but is computationally easier because no integration is required since $\mathcal{T}$ is computed via the `pt` function in base R.

The short CI uses the 2×1 Newton adjustment vector

$$\frac{1}{\sqrt{n}} \begin{bmatrix} \mathcal{T}'_L & -\mathcal{T}'_U \\ \mathcal{T}''_L & -\mathcal{T}''_U \end{bmatrix}^{-1} \begin{bmatrix} \mathcal{T}_L - \mathcal{T}_U - \lambda \\ \mathcal{T}'_L - \mathcal{T}'_U \end{bmatrix} \tag{5.22}$$

starting from the previous equal-tail CI where the subindices L and U indicate the noncentrality parameter $\sqrt{n}\theta_L$ and $\sqrt{n}\theta_U$, respectively.

Unbiased CI

The limits of the unbiased CI (3.4) are obtained by solving the system of equations

$$\mathcal{T}_L - \mathcal{T}_U - \lambda = 0, \; t_L - t_U = 0, \tag{5.23}$$

where t is the pdf of the noncentral t-distribution with $n-1$ df evaluated at $\sqrt{n}\theta_L$ and $\sqrt{n}\theta_U$, respectively. The solution for θ_L and θ_U is found iteratively by Newton's algorithm using the adjustment vector similar to the previous, with the derivative given by

$$t' = \frac{2\sqrt{n}}{\sqrt{n-1}} \int_0^\infty w^2(u(s))\phi(u(s))c_{n-1}(w^2)dw.$$

MC and MO estimators of ES

The MC estimator $\widehat{\theta}_{MC}$ solves the equation $T''(S; \sqrt{n}\theta) = 0$, and the MO estimator $\widehat{\theta}_{MO}$ solves the equation $t'(S; \sqrt{n}\theta) = 0$. Both estimates are found by Newton's algorithm starting from the sample estimate $\widehat{\theta}$; the third derivative T''' is found by differentiating T'' given by (5.21). Recall that the interval around the MC estimate covers θ with maximum probability in a small neighborhood and the MO estimate maximizes the density at the observed S. The percent difference between MC and MO estimates (y-axis) and the sample ES $\widehat{\theta}$ (x-axis) for three sample sizes are depicted in Figure 5.12. Understandably, the smaller the sample size the bigger the difference. The maximum difference is about 15%. Interestingly, the MC and MO estimates are on opposite sides of $\widehat{\theta}$.

The algorithms for optimal confidence intervals and the respective estimates are implemented in the R function sES and illustrated next with an example. The optimal CIs are compared with those from the R package MBESS by Kelley (2007).

Example 5.6 *The sample ES with $n=10$ is $\widehat{\theta} = \overline{Y}/\widehat{\sigma} = 0.25$. Compute the MC and MO estimates, equal-tail, short, and unbiased 95% CIs. Use the package MBESS for comparison.*

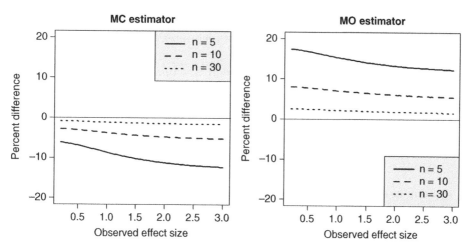

Figure 5.12: *Percent difference from the sample ES for different sample sizes.*

The output of the R function sES is shown here.

```
> sES()
Sample ES=1.5, n=10
MC estimate=1.434392, MO estimate=1.599088
95% confidence limits
            MBESS Equal-tail CI   Short CI Unbiased CI
Lower 0.5605141     0.5605142 0.5450346    0.6143691
Upper 2.4034362     2.4034362 2.3868840    2.4848073
```

The MC and MO estimates are on opposite sides of the sample ES which matches the results in Figure 5.12. The MBESS and our equal-tail CIs are identical because both use the equal-tail approach. As expected, the short and unbiased CIs are shifted in the same direction as the respective estimates.

Exact double-sided tests

The equal-tail test is straightforward: given the observed value S, the null hypothesis $H_0 : \theta = \theta_0$ versus $H_A : \theta \neq \theta_0$ is not rejected if $\alpha/2 < T(S; \sqrt{n}\theta_0) < 1 - \alpha/2$. Function T is computed in R as pt or using integrate from (5.20). The power function of all three tests is expressed in the same way,

$$P(\theta; \theta_0) = 1 + T(q_L; \sqrt{n}\theta) - T(q_U; \sqrt{n}\theta), \qquad (5.24)$$

but with different quantiles q_L and q_U depending on θ_0. For example, for the equal-tail test q_L and q_U are found by solving the equations $T(q_L; \sqrt{n}\theta_0) = \alpha/2$ and $T(q_U; \sqrt{n}\theta_0) = 1 - \alpha/2$. The computation of these quantiles is easy through a built-in R function qt. Although the size of the equal-tail test is exact it is biased because the derivative of P with respect to θ evaluated at $\theta = \theta_0$ does not vanish – see Figure 5.13 for an illustration with $\alpha = 0.05$, $n = 5$, and $\theta_0 = 2$ (the tangent line has a slight negative slope).

The unbiased exact test has the acceptance interval

$$T(q_L; \sqrt{n}\theta_0) < T(S; \sqrt{n}\theta_0) < T(q_U; \sqrt{n}\theta_0),$$

where q_L and q_U are the solutions of the system

$$T(q_U; \sqrt{n}\theta_0) - T(q_L; \sqrt{n}\theta_0) - \lambda = 0,$$
$$T'(q_L; \sqrt{n}\theta_0) - T'(q_U; \sqrt{n}\theta_0) = 0. \qquad (5.25)$$

This system is solved by Newton's algorithm with the adjustment vector for (q_L, q_U) given by

$$\frac{1}{\sqrt{n}} \begin{bmatrix} -t(q_L; \sqrt{n}\theta_0) & t(q_U; \sqrt{n}\theta_0) \\ t'(q_L; \sqrt{n}\theta_0) & -t'(q_U; \sqrt{n}\theta_0) \end{bmatrix}^{-1} \begin{bmatrix} T(\theta_U; \sqrt{n}\theta_0) - T(q_L; \sqrt{n}\theta_0) - \lambda \\ T'(q_L; \sqrt{n}\theta_0) - T'(q_U; \sqrt{n}\theta_0) \end{bmatrix}.$$

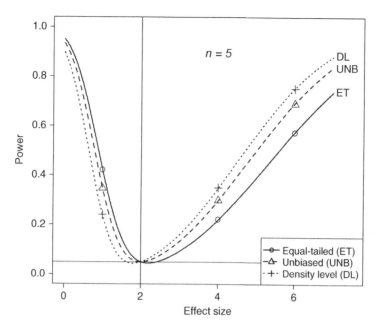

Figure 5.13: *The power function of three tests for the null hypothesis $H_0 : \theta = \theta_0$ versus $H_A : \theta \neq \theta_0$ with $\alpha = 0.05$, $n = 5$, and $\theta_0 = 2$ computed by formula (5.24). The symbols depict simulation-derived power values via the rejection rule.*

Although a built-in function `dt` for t can be used, we need the integral expressions (5.21) to compute the derivative t' given by

$$t'(s; \sqrt{n}\theta) = \frac{2\sqrt{n}}{\sqrt{n-1}} \int_0^\infty w^2(u(s))\phi(u(s))c_{n-1}(w^2)dw.$$

The density level (short acceptance interval) test has the same expression for the acceptance rule and power function but with q_L and q_U found by solving a different system of equations:

$$\mathcal{T}(q_U; \sqrt{n}\theta_0) - \mathcal{T}(q_L; \sqrt{n}\theta_0) - \lambda = 0,$$
$$t(q_L; \sqrt{n}\theta_0) - t(q_U; \sqrt{n}\theta_0) = 0. \tag{5.26}$$

Again Newton's algorithm applies with the adjustment vector given by

$$\frac{1}{\sqrt{n}} \begin{bmatrix} -t(q_L; \sqrt{n}\theta_0) & t(q_U; \sqrt{n}\theta_0) \\ t'_s(q_L; \sqrt{n}\theta_0) & -t'(\theta_U; \sqrt{n}\theta_0) \end{bmatrix}^{-1} \begin{bmatrix} \mathcal{T}(q_U; \sqrt{n}\theta_0) - \mathcal{T}(q_L; \sqrt{n}\theta_0) - \lambda \\ t(q_L; \sqrt{n}\theta_0) - t(q_U; \sqrt{n}\theta_0) \end{bmatrix},$$

where t'_s indicates differentiation with respect to the first argument, that is,

$$t'_s(s; \sqrt{n}\theta) = -\frac{2}{n-1} \int_0^\infty w^3(u(s))\phi(u(s))c_{n-1}(w^2)dw.$$

The powers of the three tests are compared graphically in Figure 5.13 for $\theta_0 = 2$, $n = 5$, and $\alpha = 0.05$. All three tests are exact, but only the UNB test is unbiased (MBESS and ET produce the same power). The symbols represent the simulation-derived power values based on the rejection rule at three alternative ES values: $\theta = 1, 4$, and 6. The match between theoretical and simulation-derived power is excellent.

Computation of the p-value

The computation of the p-value is straightforward following the discussion in Section 4.1. The unbiased and DL test depend on the target significance level α, while the equal-tail test does not. The following example, as a continuation of Example 5.6, contains the R code for computing the power of the tests and the respective p-values; see the function testES.

Example 5.7 *(a) **The null hypothesis is that the ES $\theta_0 = 0.5$.** Compute the power of detection of the alternative $\theta = 1$ having the sample size $n = 10$. (b) The sample ES with $n = 10$ is $\widehat{\theta} = 0.25$. Compute the p-values for the ET, unbiased, and DL tests.*

(a) Once q_L and q_U are determined, the powers are computed with the same formula (5.24). The power computations are carried out by calling testES(job=1). (b) The p-value computations are carried out by calling testES(job=2).

```
> testES(job=1)
ES.null=0.5, ES.alt=1 n=10
Power equal-tail test = 0.2093
Power inbiased test = 0.2326
Power density level test = 0.2516
> testES(job=2)
sample ES=0.25, ES.null=0.5 n=10
P-value equal-tail test = 0.4243
Power inbiased test = 0.5091
Power density level test = 0.615
```

The powers and p-values differ, but not drastically. Detecting the specified value of the alternative ES requires increasing the sample size to reach the accepted power value, such as 0.8 or 0.9.

5.6.2   Coefficient of variation

The coefficient of variation (CV) is one of the most important statistical parameters. Its distribution is heavily skewed and therefore M-statistics is relevant. In Section 5.3, we developed exact unbiased statistical inference for the CV for the lognormal distribution. Now we modify the algorithms developed for the ES in Section 5.6.1 to the CV.

The development of a confidence interval for CV goes back to the 1930s – it was the topic of research by several prominent statisticians including McKay (1932), Pearson (1932), and Fieller (1932), among others. The CV is an example of a statistical parameter for which the classic theory of the minimum variance unbiased estimator breaks down because the sample statistic $\widehat{\sigma}/\overline{Y}$ does not have a finite mean and an unbiased estimator for CV does not exist ($\widehat{\sigma}^2$ is the traditional unbiased estimator of variance). Indeed, as we shall learn later, the confidence interval becomes disjoint with positive probability.

Having a random sample $Y_i \overset{\text{iid}}{\sim} \mathcal{N}(\mu, \sigma^2)$, $i = 1, 2, ..., n$, we aim to develop the exact statistical inference for the population CV, $\kappa = \sigma/\mu$, reciprocal of the ES, estimated as $\widehat{\kappa} = \widehat{\sigma}/\overline{Y}$. Since M-statistics is user-friendly with respect to reparametrization, we use the same statistic $S = \sqrt{n}\overline{Y}/\widehat{\sigma}$ as in Section 5.3 for our developments with the underlying cdf $F(s; \kappa) = \mathcal{T}(s; \sqrt{n}/\kappa)$, where $\mathcal{T}$ denotes the cdf of the noncentral t-distribution with $n - 1$ df and noncentrality parameter $\sqrt{n}/\theta$.

Equal-tail CI

The equal-tail CI for $\kappa = 1/\theta$ is derived as the reciprocal of the respective CI for the ES from the previous section but not without a caveat because $\mathcal{T}(s; \sqrt{n}/\kappa)$ is discontinuous at $\kappa = 0$. Specifically, let θ_L and θ_U be the $100(1 - \alpha)\%$ equal-tail confidence limits for the ES as the solutions of the equations $\mathcal{T}(s; \sqrt{n}\theta_L) = 1 - \alpha/2$ and $\mathcal{T}(s; \sqrt{n}\theta_U) = \alpha/2$ found by Newton's algorithm as outlined in Section 5.6.1. It is tempting to take the reciprocal of the confidence limits for θ, but more care required. Let $\kappa_L = 1/\theta_U$ and $\kappa_U = 1/\theta_L$, under the assumption that $\theta_L < \theta_U$. Although it is true that $\kappa_L < \kappa_U$ if $0 < \theta_L < \theta_U$ or $\theta_L < \theta_U < 0$, the inequality $\kappa_L < \kappa_U$ does not hold if $\theta_L < 0 < \theta_U$: that is, when the lower limit is negative and the upper limit is positive. Specifically, the exact $100(1 - \alpha)\%$ CI for κ is defined as follows:

$$
\begin{aligned}
(\kappa_L, \kappa_U) \text{ when } \kappa_L \leq \kappa_U \\
(-\infty, \kappa_U) \cup (\kappa_L, \infty) \text{ when } \kappa_L > \kappa_U
\end{aligned}
\tag{5.27}
$$

Note that in the case $\kappa_L > \kappa_U$, the CI is a union of disjoint semi-infinite intervals, that is, either $\kappa < \kappa_U$ or $\kappa > \kappa_L$. Note that the definition (5.27) applies to CIs for κ as reciprocal of CIs for θ, and as such, to the short and unbiased CIs developed shortly, but it does not necessarily applies to other CIs. Figure 5.14 illustrates the simulation-derived probability of coverage of the true CV (number of simulations = 10,000). Curve **1** depicts the coverage probability of the CI defined by (5.27). Curve **2** depicts the coverage of the naive CI defined as (κ_L, κ_U). Curve **3** depicts the coverage probability of two disjoint intervals as defined in the second row of (5.27). Finally, **4** depicts the proportion of simulations for

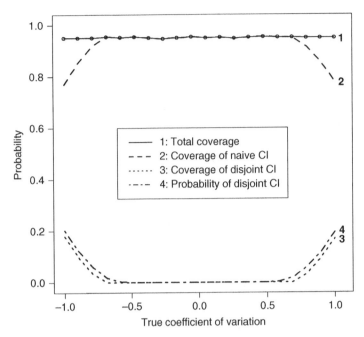

Figure 5.14: *Simulation-derived probabilities of the equal-tailed CI for the true CV, κ, with $n = 10$ and $\alpha = 0.05$.*

which $\kappa_U < \kappa_L$, leading to a disjoint CI. In short, irregularities arise when the CV is fairly large in absolute value, say, $|\kappa| > 50\%$. As we see, the total coverage probability of interval (5.27) is close to the nominal value.

Figure 5.15 depicts the simulation-derived coverage probability of six CIs (number of simulations = 2,000) with $\kappa_0 = 0.5$: our equal-tail (5.27), unbiased and short CIs (see the following section), and three previously published methods (McKay 1932; Miller 1991, Vangel 1996). The R function `CoefVarCI` from the package `cvcqv` was used to compute the latter three CIs. See the function `ciCV_comp` for details of implementation. Our CIs adjusted by the rule (5.27) are exact; McKay's, Miller's, and Vangel's coverage probabilities are below λ for small sample sizes and approach the nominal probability with n.

Short and unbiased CIs

Since the derivative of the cdf expressed through κ is

$$\frac{d}{d\kappa}\mathcal{T}(S; \sqrt{n}/\kappa) = -\frac{\sqrt{n}}{\kappa^2}\mathcal{T}(S; \sqrt{n}/\kappa) = -\sqrt{n}\theta^2\mathcal{T}(S; \sqrt{n}\theta),$$

where $\theta = 1/\kappa$, the limits of the short CI for κ can be derived from the limits for θ by solving the system

$$\mathcal{T}_L - \mathcal{T}_U - \lambda = 0, \; \theta_L^2\mathcal{T}_L' - \theta_U^2\mathcal{T}_U' = 0. \tag{5.28}$$

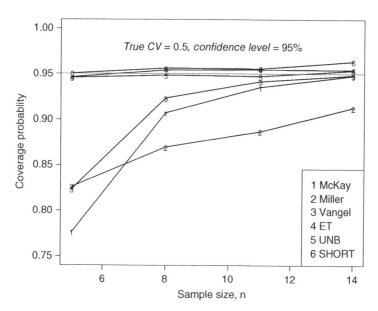

Figure 5.15: *Simulation-derived coverage probability of six CIs for the coefficient of variation (number of simulation equals 2,000). The adjustment rule (5.27) was applied to CIs 4-6. This graph was created with the function* ciCV_comp.

The confidence limits for the CV are the reciprocals: $\kappa_U = 1/\theta_L$ and $\kappa_L = 1/\theta_U$. Newton's iterations are similar to (5.22) but require simple modifications leading to

$$\frac{1}{\sqrt{n}} \begin{bmatrix} T'_L + \delta & -T'_U \\ 2\theta_L T'_L + \theta_L^2 T''(S;\theta_L) & -2\theta_U T'_U - \theta_U^2 T''_U + \delta \end{bmatrix}^{-1} \begin{bmatrix} T_L - T_U - \lambda \\ \theta_L^2 T'_L - \theta_U^2 T'_U \end{bmatrix},$$

where $\delta > 0$ is the regularization parameter. We found that $\delta = 10^{-3}$ works well although the iterations may take longer. After θ_L and θ_U are determined, the rule (5.27) applies.

Thanks to the invariance property of the MO-statistics, the unbiased CI for the CV is reciprocal of ES, that is, the CI for κ is defined as $(1/\theta_L, 1/\theta_U)$, where θ_L and θ_U are derived from (5.23).

MC and MO estimators of CV

According to the general definition 2.5, the MC estimator is the limit point of the short CI when $\lambda \to 0$. First, we find this limit point, $\widehat{\theta}_{MC}$, from (5.28); second, let $\widehat{\kappa}_{MC} = 1/\widehat{\theta}_{MC}$. By definition, $\theta = \widehat{\theta}_{MC}$ is defined from

$$\lim_{\theta_L \to \theta, \theta_U \to \theta} \frac{\theta_L^2 T'(S; \sqrt{n}\theta_L) - \theta_U^2 T'(S; \sqrt{n}\theta_U)}{\theta_L - \theta_U} = 0,$$

or equivalently, by solving the equation

$$2T'(S; \sqrt{n}\theta) + \sqrt{n}\theta T''(S; \sqrt{n}\theta) = 0. \tag{5.29}$$

Newton's algorithm takes the form

$$\theta_{s+1} = \theta_s - \frac{2n^{-1/2}T'(S; \sqrt{n}\theta_s) + \theta_s T''(S; \sqrt{n}\theta_s)}{(2 + \theta_s)T''(S; \sqrt{n}\theta_s) + \theta_s T'''(S; \sqrt{n}\theta_s)}, \quad s = 0, 1, \ldots$$

starting from $\theta_0 = 1/\widehat{\kappa}$. Adding a regularization parameter to the denominator, say, $\delta = 0.01$ may improve the convergence (an example of the function ciCV is presented shortly).

Note that when n is large the second term at the left-hand side of (5.29) dominates. This means for large n, the MC estimator of the CV is the reciprocal of the MC estimator of the ES. Due to the invariance to reparametrization, we have $\widehat{\kappa}_{MO} = 1/\widehat{\theta}_{MO}$. This result is seen directly from the differentiation of the density with respect to κ:

$$\frac{d}{d\kappa}t(S; \sqrt{n}/\kappa) = -\frac{\sqrt{n}}{\kappa^2}t'(S; \sqrt{n}\theta).$$

This means the left-hand side equals zero when $\sqrt{n}/\kappa = \sqrt{n}\theta$, that is, $\widehat{\kappa}_{MO} = 1/\widehat{\theta}_{MO}$. Figure 5.16 shows the two estimates for three sample sizes as a function of the observed $\widehat{\kappa}$ on the percent scale.

The following example is similar to Example 5.6: it contains the R code for all CI algorithms discussed earlier.

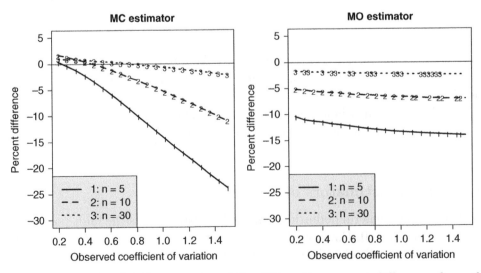

Figure 5.16: *MC and MO estimates of the CV as the percent difference from the observed/sample CV. In general, both estimators suggest an increased value of* $\widehat{\kappa}$.

Example 5.8 *The sample CV with* $n = 10$ *is* $\hat{\kappa} = \hat{\sigma}/\overline{Y} = 0.5$. *Compute the MC and MO estimates,the equal-tail, short, and unbiased 95% CIs.*

The R function `ciCV` contains all the algorithms. The call and the output are shown next.

```
> ciCV()
Sample CV=0.5, n=10
MC estimate=0.4984881, MO estimate=0.4706225
95% confidence limits
        Equal-tail CI    Short CI  Unbiased CI
Lower      0.3243264   0.2816532    0.3136103
Upper      1.1313480   0.9730693    1.0537390
ET longer than SHORT by 17%
UNB longer than SHORT by 7%
```

The MC and MO estimates of the CV are smaller than the sample CV $= 0.5$ in agreement with Figure 5.16. Both the short and unbiased CIs are shifted downward. The lengths of the ET and the unbiased CI are greater than the length of the short CI by 17% and 7%, respectively.

5.6.3 Double-sided hypothesis tests

The three tests developed for the ES are easily modified to the CV with the null hypothesis $H_0 : \kappa = \kappa_0$ versus the alternative $H_A : \kappa \neq \kappa_0$. All the tests have the same expression for the power function

$$P(\kappa; \kappa_0) = 1 + \mathcal{T}(1/q_U; \sqrt{n}/\kappa) - \mathcal{T}(1/q_L; \sqrt{n}/\kappa),$$

where q_L and q_U depend on κ_0 but vary from test to test. For all three tests, the hypothesis is accepted if the observed $S = \sqrt{n}\overline{Y}/\hat{\sigma}$ falls within the interval (q_L, q_U).

Now we define q_L and q_U for all three tests. For the equal-tail test q_L and q_U are defined from $\mathcal{T}(q_L; \sqrt{n}/\kappa_0) = \alpha/2$ and $\mathcal{T}(q_U; \sqrt{n}/\kappa_0) = 1 - \alpha/2$ as the respective quantiles of the noncentral t-distribution (the `qt` function in R).

The density level test requires solving the system for q_L and q_U, the same as for the ES given by (5.26) where $\theta_0 = 1/\kappa_0$. The same holds for the unbiased test with the system given by (5.24) since the ES and CV are reciprocal.

The three power functions are portrayed in Figure 5.17, and created with the R function `ncv` – this function contains codes for all three algorithms. The difference to the left of the null value $\kappa_0 = 0.5$ is small but to the right is quite visible due to the asymmetry of the noncentral distribution. The numbers, corresponding to the three tests, depict the simulation-derived power values. The numbers at $\kappa = 0.75$ depict the proportion of simulations for which S falls outside the interval (q_L, q_U), the rejection rule, and the numbers at $\kappa = 1$ depict

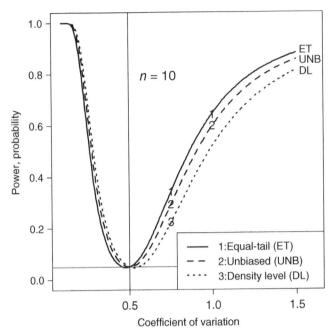

Figure 5.17: *Three power functions for testing the double-sided null hypothesis* $H_0 : \kappa = 0.5$. *The numbers 1-3 at $\kappa = 0.75$ depict simulation derived power values according to the rejection rule, and the numbers at $\kappa = 1$ depict the power values according to the dual CIs (number of simulations = 10,000). This graph was created by issuing* **ncv** *( )*.

the proportion of simulations for which the dual CI covers κ_0 (the dual CI is computed only for the ET and UNB tests).

5.6.4　Multivariate ES

Statistical inference for the univariate ES/standardized mean and its reciprocal, the multivariate CV, can be generalized to the multivariate case with the t-distribution replaced with F-distribution. The multivariate CV is frequently used for monitoring the quality control of technological processes in control charts (Guo and Wang 2018). In this section, we will discuss optimal statistical inference for the multivariate ES (MES) due to mathematical convenience.

The m-dimensional vectors of observations follow a multivariate normal distribution

$$\mathbf{y}_i \overset{\text{iid}}{\sim} \mathcal{N}(\boldsymbol{\mu}, \boldsymbol{\Omega}), \; i = 1, 2, ..., n.$$

We are concerned with the statistical inference of the population MES defined as

$$\theta = \boldsymbol{\mu}' \boldsymbol{\Omega}^{-1} \boldsymbol{\mu}$$

using its sample counterpart

$$\widehat{\theta} = \overline{\mathbf{y}}'\widehat{\boldsymbol{\Omega}}^{-1}\overline{\mathbf{y}},$$

where $\widehat{\boldsymbol{\Omega}}$ is the unbiased estimator of $\boldsymbol{\Omega}$. Statistical inference for the MES relies on the fact that its distribution is expressed through the F-distribution as follows:

$$\Pr(\widehat{\theta} \le x) = \mathcal{F}\left(\frac{n(n-m)}{(n-1)m}x; m, n-m, \mathrm{ncp} = n\theta\right), \, x > 0.$$

where $\mathcal{F}$ is the cdf of the F-distribution. M-statistics applies to statistical inference of θ straightforwardly. Section 5.12 provides some guidelines for treating the noncentrality parameter in the F-distribution.

5.7   Binomial probability

Estimation of binomial probability from iid Bernoulli experiments is one of the oldest statistical problems. Yet, the traditional unbiased estimator $\widehat{p} = m/n$, where m is the number of successes and n is the number of trials, is not without glitches: the probability parameter, p, is assumed positive, but with small n, the chance that $m = 0$ cannot be ignored especially for small p that yields $\widehat{p} = 0$.

To illustrate the paradox, consider the estimation of the probability of getting into a car accident for a specific individual on a particular day. Everybody agrees that this probability is positive, although tiny. For an individual who drove almost every day during a specific year, but did not get into a car accident, the unbiased estimate of the probability is zero which contradicts common sense. The MC-statistics solves this problem by abandoning unbiased estimation. In particular, we aim to suggest and theoretically justify a nonzero estimator of the probability for rare events when $m = 0$ is quite probable.

Classic unbiased estimation of the binomial probability leads to another paradox that contradicts common sense: Team 1 played two matches and won both of them. Team 2 played 100 matches and won 99. Which team is better? Using the classic approach, the first probability estimate is $2/2 = 1$, and the second probability estimate is $99/100$. This concludes that the first team is better, but again contradicts common sense: everybody picks the second team as being better.

Before moving to statistical inference for binomial probability we want to make an important comment on replacing the original discrete distribution with a continuous one on the range of discrete outcomes. Technical difficulties associated with the discontinuity of the cdf make it impossible to developing CIs with constant coverage probability and hypothesis tests with exact type I error. Several authors abandoned the "exact" approach and argued that approximated statistical inference for discrete distribution is better, nicely illustrated by Agresti and Coul (1998). The problem of demanding the exact coverage or the size of the test was recognized many years ago with a suggestion of randomization of the rejection decision (Lehmann and Casella 1998). Instead of randomization, we propose replacing the original discontinuous cdf with its continuous counterpart. Indeed, this method has been known for many years and applied by Clopper and Pearson (1934) by replacing the original discrete binomial cdf with the beta cdf to develop an "exact" equal-tail CI for the binomial probability. However, we propose a small adjustment to the shape of the beta distribution that makes the statistical inference symmetric with respect to the complementary probability. Although the replacement of a discrete with a continuous distribution makes computations easy it does not eliminate the fundamental difficulty – the coverage probability is not exact and has a saw-tooth shape when plotted against the true p.

5.7.1   The MCL estimator

To avoid summation and enable the derivation with respect to the outcome we express the binomial cdf via the cdf of the beta distribution. Then the binomial cdf, $F_n(m; p)$, where n is the number of Bernoulli trials and m is the number of successes, is written as $F_n(m; p) = 1 - \mathcal{B}(p; m + 1, n - m)$, where $\mathcal{B}$ stands for the cdf of the beta distribution with the shape parameters $m + 1$ and $n - m$. This representation of the binomial cdf through the beta cdf hereafter is called Clopper-Pearson's representation.

Note that this representation is not symmetric with respect to relabeling "success" and "failure" or formally, $F_n(0; p) \neq 1 - F_n(n; 1 - p)$. To avoid this unwanted asymmetry we propose the following approximation with the beta distribution, hereafter referred to as the beta21 approximation:

$$F_n(m; p) \simeq \tilde{F}_n(m; p) \stackrel{\text{def}}{=} 1 - \mathcal{B}(p; m + 1, n - m + 1)$$

$$= \frac{1}{B(m + 1, n - m + 1)} \int_p^1 u^m (1 - u)^{n-m} du, \qquad (5.30)$$

where B is the complete beta function and $m = 0, 1, ..., n$. The beta21 approximation is the same as the beta representation of the binomial distribution used by Clopper and Pearson but with an additional 1 at the first shape parameter. Here 2 indicates that the second shape parameter is adjusted and 1 indicates the adjustment. Note that unlike the original binomial distribution, $\tilde{F}_n(m; p)$ is defined for continuous $0 \leq m \leq n$.

Proposition 5.9 *The beta21 approximation of the binomial cdf is symmetric with respect to relabeling "success" and "failure," that is, $\tilde{F}_n(0; p) = 1 - \tilde{F}_n(n; 1 - p)$ for any p.*

Proof. Since

$$\tilde{F}_n(0;p) = 1 - \mathcal{B}(p;1,n+1), \ \tilde{F}_n(n;1-p) = 1 - \mathcal{B}(1-p;n+1,1)$$

it is sufficient to prove that

$$\mathcal{B}(p;1,n+1) = 1 - \mathcal{B}(1-p;n+1,1).$$

This identity may be viewed as a special case of the general symmetry of the beta cdf, $\mathcal{B}(x;j,k) = 1 - \mathcal{B}(1-x;k,j)$, sometimes referred to as reflection symmetry plus unitary translation, but we prove the symmetry anyway. Find

$$\mathcal{B}(p;1,n+1) + \mathcal{B}(1-p;n+1,1)$$

$$= \frac{1}{B(1,n+1)} \int_0^p (1-u)^n du + \frac{1}{B(n+1,1)} \int_0^{1-p} u^n du. \qquad (5.31)$$

Notice that

$$B(1,n+1) = \frac{\Gamma(n+2)}{\Gamma(1)\Gamma(n+1)} = B(n+1,1).$$

Hence, after a change of variable $v = 1 - u$, rewrite (5.31) as

$$\frac{1}{B(1,n+1)} \left[\int_0^{1-p} u^n du + \int_{1-p}^1 v^n dv \right] = 1.$$

This proves the beta21's symmetry. $\qquad \square$

The original binomial cdf and its beta approximation as functions of m and p are shown in Figure 5.18 for $n = 8$ and $m = 3$. Note that the Clopper-Pearson

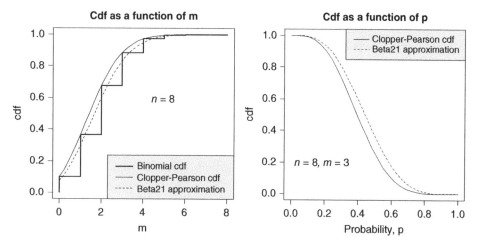

Figure 5.18: *Cdf of the binomial distribution and its approximations as functions of the number of successes (m) and the probability parameter (p).*

beta cdf takes the same values at discrete values m as the original cdf, as indicated previously, but the beta21 cdf approximation does not, and is shifted down.

An immediate advantage of the approximation (5.30) is that the equal-tail $100\lambda\%$ CI for p can be easily computed as $p_L = \mathcal{B}^{-1}((1-\lambda)/2; m+1, n-m+1)$ and $p_U = \mathcal{B}^{-1}((1+\lambda)/2; m+1, n-m+1)$, where $\mathcal{B}^{-1}$ denotes the quantile beta (qbeta in R). This method will be referred to as the modified equal-tail (MET) CI (it is different from Clopper-Pearson's because of the adjusted df). Note that p_L and p_U, based on approximation (5.30) exist for every m including $m = n$, but the equal-tail CI does not exist when $m = n$ because $F_n(n; p) = 1 - \mathcal{B}(p; n+1, 0) = 1$ regardless of p. This implies that for the classic Clopper-Pearson CI

$$\mathcal{B}^{-1}(\alpha/2; m, n-m+1) < p < \mathcal{B}^{-1}(1-\alpha/2; m+1, n-m).$$

We have $p_L = 0$ when $m = 0$ and $p_U = 1$ when $m = n$, while the MET CI stays within $(0, 1)$.

5.7.2   The MCL2 estimator

The goal of this section is to derive the maximum concentration estimator of $p \in (0, 1)$ that bounds from extremes 0 and 1 even when the number of observed events is either 0 or n. To avoid 0 or 1 estimates even when $m = 0$ or $m = n$, the double-log scale is applied. Recall that the MCL2 estimator for p is derived as the limit of the CI when $\lambda \to 0$ with the length of the interval measured on the double-log scale as outlined in Section 2.5. That is if p_L and p_U are the lower and upper limit of the CI ($0 < p_L < p_U < 1$) then the "length" of the interval is measured as

$$(\ln p_U - \ln p_L) + (\ln(1 - p_L) - \ln(1 - p_U)).$$

Equivalently, it solves equation (2.17) or specifically,

$$(1 - 2p)\tilde{F}'_n(m; p) + p(1 - p)\tilde{F}''_n(m; p) = 0,$$

where

$$\tilde{F}'_n(m; p) = -\frac{1}{B(m+1, n-m+1)} p^m (1 - p)^{n-m}, \tag{5.32}$$

the negative pdf of the beta distribution (dbeta) with the shape parameters $\alpha = m + 1$ and $\beta = n - m + 1$. Note that both cdf and pdf of the beta distribution are readily available in R (pbeta and dbeta, respectively).

After omitting the parameter-independent term $B(m+1, n-m+1)$, we arrive at the equation

$$(1 - 2p)p^m (1 - p)^{n-m} + p(1 - p)(mp^{m-1}(1 - p)^{n-m}$$

$$-(n - m)p^m (1 - p)^{n-m-1}) = 0.$$

Finally, some algebra yields the sought MCL2 estimator of the binomial probability:

$$\widehat{p}_{MCL2} = \frac{m+1}{n+2},$$

known as Laplace's law of succession (Wilson 1927). Note that even for $m = 0$, the MCL2 estimator is not zero but $1/(n+2)$. This estimator is invariant with respect to relabeling "success" and "failure" meaning $\widehat{p}_{MC2} = 1 - \widehat{\theta}_{MCL2}$ where $\theta = 1 - p$ is the probability of failure, or algebraically

$$\frac{m+1}{n+2} = 1 - \frac{(n-m)+1}{n+2}.$$

This means when $m = n$, we have $\widehat{p}_{MCL2} = 1 - 1/(n+2) = (n+1)/(n+2)$.

In our first example, even if the driver did not get into a single car accident during one year ($m = 0$), the probability that they will get into a car accident on a particular day is $\widehat{p}_{MCL2} = 1/(365+2) = 0.0027$.

In our second example with the two teams, the second team is better according to the MCL2 estimator because the probability of winning in a single game for the first team is $(2+1)/(2+2) = 3/4$ and for the second team is $(99+1)/(100+2) = 100/102$. This makes sense now.

5.7.3   The MCL2 estimator of p^n

The fact that there are no unbiased sensible estimators of p^2 is well known. This fact is easy to generalize to p^n, the probability that all n Bernoulli experiments are successes ($n \geq 2$). We aim to prove that there are no sensible unbiased estimators of p^n. Indeed, if $\widehat{\theta}$ is an unbiased estimator of p^n, then

$$\sum_{m=0}^{n} r_m p^m (1-p)^{n-m} - p^n = 0 \quad \forall p \in (0,1),$$

where $r_m = t_m \binom{n}{m}$ and $\{t_m, m = 0, 1, ..., n\}$ are the nonnegative values $\widehat{\theta}$ takes. Specifically, we aim to prove that the only sequence r_m that makes the left-hand side zero is when $r_m = 0$ for $m < n$ and $r_n = 1$. Collect the coefficients at p^m. By direct examination we can show that the coefficient at p^0 is r_0 and the coefficient at p^m is $\sum_{k=0}^{m} c_k r_k$, which is a linear combination of $\{r_k, k = 0, ..., m\}$. Moving from $m = 0$ to n, we find that $r_0 = 0, r_1 = 0, ..., r_{n-1} = 0$, and therefore $r_n = 1$, which is was we intended to prove. This means the only unbiased estimator $\widehat{\theta}$ of p^m is such that it takes a zero value for all $m < n$ and $\widehat{\theta} = 1$ if $m = n$, an unacceptable estimator. $\qquad\square$

Now we turn our attention to the MC estimator on the double-log scale because $0 < p^n < 1$. Express p through the parameter of interest $\theta = p^n$ as $p = \theta^{1/n}$, and express $\tilde{F}_n$ through θ as follows:

$$\tilde{F}_n(m; \theta) = \frac{1}{B(m+1, n-m+1)} \int_{\theta^{1/n}}^{1} u^m (1-u)^{n-m} du.$$

Instead of solving equation (2.17), minimize $\theta(1-\theta)\tilde{F}'_n(m;\theta)$, which is equivalent to maximization of

$$(1-\theta)\theta^{(m+1)/n}(1-\theta^{1/n})^{n-m}. \tag{5.33}$$

We reduce this optimization problem to maximization of the polynomial $p^{m+1}(1-p^n)(1-p)^{n-m}$ by letting $p = \theta^{1/n}$. Simple algebra shows that the maximizer $p \in (0,1)$ must satisfy the polynomial equation

$$(2n+1)p^{n+1} - (n+m+1)p^n - (n+1)p + (m+1) = 0. \tag{5.34}$$

An advantage of solving a polynomial equation over the maximization of (5.33) is that all real roots on the interval $(0,1)$ can be found (function polyroot in R). It's easy to see that $p = 1$ is a root because

$$(2n+1) - (n+m+1) - (n+1) + (m+1) = 0.$$

Once the root p is found, the MCL2 estimator is p^n.

To illustrate, consider $n = 2$. Then the polynomial (5.34) turns into a cubic polynomial

$$5p^3 - (3+m)p^2 - 3p + m + 1 = (p-1)(5p^2 + (2-m)p - 1 - m).$$

Since $p \neq 1$, the problem of finding the roots reduces to a quadratic equation $5p^2 + (2-m)p - 1 - m = 0$, with a positive root given by

$$p_+ = \frac{1}{10}\left(\sqrt{m^2 + 16m + 24} + (m-2)\right).$$

It is easy to show that $0 < p_+ < 1$. Indeed, since $0 \leq m \leq 2$, we have

$$p_+ \leq \frac{1}{10}(\sqrt{2^2 + 16 \times 2 + 24} + (2-2)) = \frac{\sqrt{60}}{10} < 1,$$

$$p_+ \geq \frac{1}{10}(\sqrt{0^2 + 16 \times 0 + 24} + (0-2)) = \frac{\sqrt{24}-2}{10} > 0.$$

Finally, the MCL2 estimator of p^2 is p_+^2.

5.7.4 Confidence interval on the double-log scale

The problem of the CI for a single proportion is one of the oldest in statistics. Newcombe (1998) compares seven methods of interval estimation. The goal of this section is to develop a CI with lower and upper limits staying away from 0 and 1, respectively. This interval is compared to other popular methods of interval estimation.

The MCL2 CI for p is found by solving the system (2.16). For binomial probability, the second equation has the form

$$p_L^{m+1}(1 - p_L)^{n-m+1} - p_U^{m+1}(1 - p_U)^{n-m+1} = 0,$$

which upon taking the log, simplifies to

$$(m + 1)\ln(p_U/p_L) - (n - m + 1)\ln((1 - p_L)/(1 - p_U)) = 0.$$

The MCL2 CI (p_L, p_U) is the solution of this equation coupled with the coverage probability condition

$$\mathcal{B}(p_U; m + 1, n - m + 1) - \mathcal{B}(p_L; m + 1, n - m + 1) - \lambda = 0.$$

Newton's algorithm suggests iterations with the adjustment vector for $(p_L, p_U)'$ as

$$\begin{bmatrix} F_n'(m; p_L) & -F_n'(m; p_U) \\ -\frac{m+1}{p_L} + \frac{n-m+1}{1-p_L} & \frac{m+1}{p_U} - \frac{n-m+1}{1-p_U} \end{bmatrix}^{-1} \begin{bmatrix} F_n(m; p_U) - F_n(m; p_L) - \lambda \\ (m + 1)\ln\frac{p_U}{p_L} - (n - m + 1)\ln\frac{1-p_L}{1-p_U} \end{bmatrix}$$

starting from the MET CI. The equal- and unequal-tail CIs are compared graphically in Figure 5.19 for $n = 10$ and $\lambda = 0.95$. The coverage probability of both CIs has the familiar saw-toothed shape but the expected log length, as the average of $\ln p_U/p_L + \ln(1 - p_L)/(1 - p_L)$, is slightly but uniformly smaller for the MCL2 CI than the MET CI. Note that the asymptotic/Wald CI overcovers the true p.

The R code for the following example implements all algorithms discussed earlier.

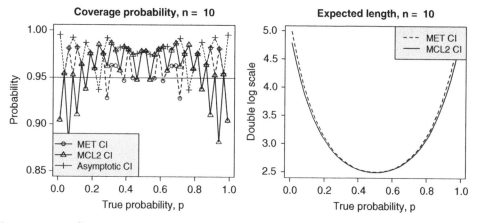

Figure 5.19: *Comparison of CIs for a binomial probability. Top: The coverage probability has the familiar saw shape due to the discreteness of the binomial distribution. Bottom: The expected length of the MCL2 CI on the double log scale is smaller for all p.*

Example 5.10 *In* $n = 10$ *Bernoulli trials* $m = 2$ *outcomes were positive.* *Compute the 95% CI for the probability of a positive outcome, p.*

See the R code `pbinCI` with the output listed here.

```
> pbinCI()
Estimation of binomial probabiliy n=10, m=2
95% confidence intervals
        Asymptotic/Wald  Agresti-Coul Clopper-Pearson      MET      MCL2
Lower       -0.04791801   -0.06018422      0.04183614 0.06021773 0.06563364
Upper        0.44791801    0.46018422      0.55609546 0.51775585 0.53602035
```

Five CIs are computed. The lower limits of the asymptotic/Wald and Agresti-Coul CI are negative. The MET and MCL2 methods shift the interval upward.

5.7.5 Equal-tail and unbiased tests

This section discusses statistical tests for the null hypothesis $H_0 : p = p_0$ versus $H_A : p \neq p_0$, sometimes referred to as the one-sample proportion test. All the tests are reduced to finding quantiles $0 \leq q_L < q_U \leq 1$, dependent on p_0 and the desired size of the test α, such that the null hypothesis is rejected if $\widehat{p} = m/n$ falls outside of the interval $(q_L, q_U]$ or, equivalently, if $\widehat{p} \leq q_L$ or $\widehat{p} > q_U$. It is straightforward to find the equal-tail binomial quantiles in R:

$$\text{qL.BIN} = \text{qbinom}(\text{alpha}/2, \text{size} = n, \text{prob} = p0)/n,$$

$$\text{qU.BIN} = \text{qbinom}(1 - \text{alpha}/2, \text{size} = n, \text{prob} = p0)/n.$$

This test is implemented in the R built-in function `binom.test`; we present it here for completeness and comparison with other tests. The Clopper-Pearson equal-tail quantiles are found via the beta distribution quantiles as

$$\text{qL.CP} = \text{qbeta}(\text{alpha}/2, \text{p0*n}, (1-\text{p0}) * n + 1),$$

$$\text{qU.CP} = \text{qbeta}(1 - \text{alpha}/2, \text{p0*n+1}, (1-\text{p0}) * n).$$

The power function of all the tests has the familiar form

$$P(p; p_0) = 1 - F_n(q_U; p) + F_n(q_L; p), \tag{5.35}$$

where F_n stands for the original discrete cdf (`pbinom`).

Now we discuss how to find the quantiles q_L and q_U when the cdf is approximated by its continuous counterpart $\tilde{F}_n$, as presented in (5.30). The equal-tail quantiles are a straightforward modification of the Clopper-Pearson quantiles by adjusting the degrees of freedom:

$$\texttt{qL.MET = qbeta(alpha/2,p0 * n + 1,(1 - p0) * n + 1)},$$

$$\texttt{qU.MET = qbeta(1 - alpha/2,p0 * n + 1,(1 - p0) * n + 1)}.$$

Now we discuss the computation of the quantiles for the unbiased test. The equations (3.1) turn into

$$\tilde{F}_n(q_U; p_0) - \tilde{F}_n(q_L; p_0) = 1 - \alpha, \ \tilde{F}'_n(q_L; p_0)) - \tilde{F}'_n(q_U; p_0) = 0.$$

From a computational perspective it is easier to work with the log transformed $\tilde{F}'_n$. Introduce the function

$$h(q) = \psi(n+2) - \psi(nq+1) - \psi(n(1-q)+1)$$
$$+ nq \ln p_0 + n(1-q) \ln(1-p_0)$$

where $\psi(x) = \ln \Gamma(x)$ stands for the log gamma function (lgamma) with its derivative

$$h'(q) = -n\psi'(nq+1) + n\psi(n(1-q)+1) + n \ln p_0 - n \ln(1-p_0),$$

where ψ' is the derivative (digamma). Then the second equation of the previous system can be rewritten as $h(q_L) - h(q_U)$. Newton's iterations have the familiar format

$$\begin{bmatrix} -\tilde{F}'_n(q_L; p_0)) & \tilde{F}'_n(q_U; p_0)) \\ h'(q_L) & -h'(q_U) \end{bmatrix}^{-1} \begin{bmatrix} \tilde{F}_n(q_U; p_0) - \tilde{F}_n(q_L; p_0) - \lambda \\ h(q_L) - h(q_U) \end{bmatrix}$$

starting from the equal-tail solutions with $\tilde{F}'_n$ given by (5.32).

Four power functions for two pairs of the sample size and null probability are depicted in Figure 5.20 (this graph was created with the R function binTEST). Symbols depict the simulation-derived power values at few alternative probability values as the proportion of simulations for which $\hat{p}$ falls outside of the interval $(q_L, q_U]$. Due to the discreteness of the binomial distribution we do not expect the power curve to go through the point (p_0, α) or the tangent line to be horizontal at p_0.

The following example illustrates the computation of the p-value.

Example 5.11 (a) The median family income in the state is $70,000. Of 15 randomly asked individuals from a town, 5 said that their family income is less than $70,000. Compute the p-value for each of the four tests displayed in Figure 5.20 to test that whether 50% families in the town have income less than $70,000, assuming that the size of the test $\alpha = 0.05$. (b) Use simulations to confirm the p-values via a theoretical power function and the rejection rule. (c) Determine how many individuals must be asked to reject the null hypothesis that 50% of the

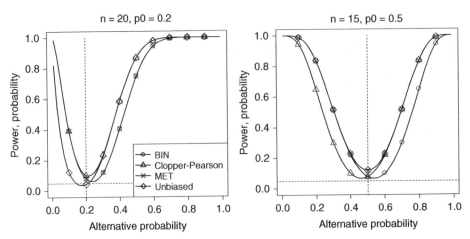

Figure 5.20: *Power functions and simulation-derived values via rejection rule for two pairs of n and p, depicted by symbols. This graph was created by function* binTEST.

town's families earn less than $70,000 versus the alternative that only 40% of families have income less than $70,000, under the assumption that the error of rejection of the null hypothesis is 5% and the error of rejection of the alternative is 20% (the desired power = 80%).

See the R function binTEST.

(a) As follows from Section 4.1 the *p*-value for the equal-tail test is given by

$$p = 2 \times \min(F(m; p_0), 1 - F(m; p_0)),$$

where $p_0 = 0.5$, $m = 5$, $n = 15$, and F is the binomial cdf. For the unbiased test, we use formula (4.3). The call binTEST(job=2) carries out the computation for task (a), and binTEST(job=3) does the computation for task (b), with the following outputs. The *p*-values are quite different.

```
> binTEST(job=2)
n=15, m=5, p0=0.5
                    P-value
Binomial          0.3017578
Clopper-Pearson   0.2748140
MET               0.1631389
Unbiased          0.1273570


> binTEST(job=3)
n=15, p0=0.5, p.alt=0.6
              P-value Theoretical Rejection rule
Binomial      0.218419  0.09984956       0.100012
```

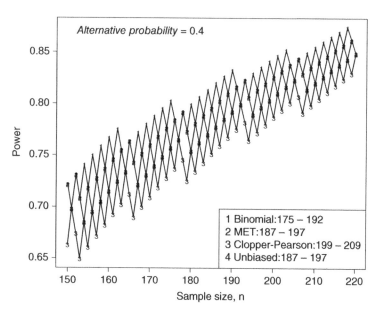

Figure 5.21: *Four power functions as functions of n, the number of individuals (sample size) to determine the interval of n for which the power is around 80%. This graph was created with the call* `binTEST(job=4)`.

Clopper-Pearson	0.218419	0.21920547	0.218419
MET	0.225805	0.22662537	0.225805
Unbiased	0.225805	0.22662537	0.225805

This computation confirms the advantage of replacing or approximating the binomial with a continuous beta distribution: the power function p-value from MET, Clopper-Pearson, and unbiased is not confirmed with the rejection rule simulation. This implies that the sample size determination based on the original binomial cdf may not produce the desired power (discussed shortly). All other p-values and simulations match.

Figure 5.21 depicts the power of four tests as a function of the number of individuals to ask, n, the sample size. Note the saw-tooth shape of the power. Since the power curve intersects the horizontal line many times we compute the minimum and maximum of the required n shown in the legend. The MET and unbiased tests produce the same intervals because the power is symmetric at $p_0 = 0.5$.

5.8 Poisson rate

The plan is to develop a novel estimator, CI, and hypothesis test for the Poisson rate using methods of M-statistics outlined in the previous chapters. First, we derive a double-sided short CI on the log scale that bounds the lower confidence limit from zero. Second, we suggest an unbiased hypothesis test. Third, we derive the MCL estimator for the Poisson rate. Similarly to the binomial probability, the MCL estimator is always positive even when all observations are zero. A note regarding the notation: since λ is traditionally used to denote the rate of the Poisson distribution, we use $1 - \alpha$ when referring to the confidence level in this section.

Theorem 5.12 *The cdf of the Poisson distribution with the rate parameter $\lambda > 0$ satisfies Assumptions 1 and 2 from Section 2.1 and 3 and 4 from Section 2.5.1.*

Proof. Assumption (1) obviously holds. (2) The probability mass function (pmf) of the Poisson distribution with the rate parameter $\lambda > 0$ is given by

$$\Pr(X = k; \lambda) = \frac{1}{k!} e^{k \ln \lambda - \lambda}, \quad k = 0, 1, 2, \dots. \tag{5.36}$$

Now we prove that the Poisson cdf given by

$$F(S; \lambda) = \sum_{k=0}^{S} \frac{1}{k!} e^{k \ln \lambda - \lambda} \tag{5.37}$$

monotonically decreases with λ. Denote the cdf evaluated at S and $S - 1$ as F_S and F_{S-1}, respectively. Find the derivative of the cdf with respect to λ, and express it through F_S and F_{S-1} as follows:

$$\frac{d}{d\lambda} \sum_{k=0}^{S} \frac{1}{k!} e^{k \ln \lambda - \lambda} = \frac{d}{d\lambda} \left[e^{-\lambda} \left(1 + \sum_{k=1}^{S} \frac{1}{k!} \lambda^k \right) \right]$$

$$= -F_S + e^{-\lambda} \sum_{k=1}^{S} \frac{1}{k!} k \lambda^{k-1} = -F_S + e^{-\lambda} \sum_{k=1}^{S} \frac{1}{(k-1)!} \lambda^{k-1}$$

$$= -(F_S - F_{S-1}) < 0.$$

(3) will be discussed later. (4)

$$\lim_{\lambda \to 0} F(S; \lambda) = \lim_{\lambda \to 0} e^{-\lambda} \left(1 + \sum_{k=1}^{S} \frac{1}{k!} \lambda^k \right) = 1 + \lim_{\lambda \to 0} \sum_{k=1}^{S} \frac{1}{k!} \lambda^k = 1.$$

The fact that $\lim_{\lambda \to \infty} F(S; \lambda) = 0$ follows from $k \ln \lambda - \lambda \to -\infty$. $\square$

5.8.1   Two-sided short CI on the log scale

We aim to develop the short CI for λ on the log scale having a random sample $\{Y_i, i = 1, \dots, n\}$ from a Poisson distribution with the rate parameter $\lambda > 0$. As an obvious statistic, we use the sum of outcomes denoted as $S = \sum_{i=1}^{n} Y_i$. It is well known that S has a Poisson distribution with the rate $n\lambda$. To facilitate the derivation, the chi-square distribution is used to express the cdf of S as follows:

$$F(s; \lambda) = 1 - C_{2(s+1)}(2n\lambda), \ s = 0, 1, 2, \ldots \tag{5.38}$$

We shall refer to this approximation as the "chi-square" representation of the cdf. An advantage of the continuous cdf representation is that s, as a part of df, can be treated as continuous, and therefore the issues related to noncompliance with having the exact size of α or continuity of the power function disappear.

The equal-tail CI limits are easy to derive based on this representation as

$$\lambda_U = (2n)^{-1} C_{2(S+1)}^{-1}(1 - \alpha/2), \quad \lambda_L = (2n)^{-1} C_{2(S+1)}^{-1}(\alpha/2),$$

where S is the observed sum and C^{-1} denotes the quantile chi-square function (the `qchisq` function in R).

To find the short two-sided CI for λ on the log scale evaluate the derivative F' up to the terms that do not depend on λ:

$$\ln(-F'(s; \lambda)) = \ln(2n \times c_{2(s+1)}(2n\lambda)) = s \ln \lambda - n\lambda.$$

The general equations (2.15) for computing the short CI on the log scale for the Poisson rate is rewritten as

$$C_{2(S+1)}(2n\lambda_U) - C_{2(S+1)}(2n\lambda_L) - (1 - \alpha) = 0,$$
$$(S + 1)(\ln \lambda_U - \ln \lambda_L) - n(\lambda_U - \lambda_L) = 0. \tag{5.39}$$

This system is solved iteratively by Newton's algorithm with the 2×1 adjustment vector

$$\begin{bmatrix} -2nc_{2(S+1)}(2n\lambda_U) & 2nc_{2(S+1)}(2n\lambda_L) \\ -(S+1)/\lambda_L + n & (S+1)/\lambda_U - n \end{bmatrix}^{-1}$$
$$\times \begin{bmatrix} C_{2(S+1)}(2n\lambda_U) - C_{2(S+1)}(2n\lambda_L) - (1 - \alpha) \\ (S+1)(\ln \lambda_U - \ln \lambda_L) - n(\lambda_U - \lambda_L) \end{bmatrix} \tag{5.40}$$

starting from the equal-tail CI given previously. Note that the traditional (additive scale) optimal CI would minimize $\lambda_U - \lambda_L$ with S instead of $S+1$ in the second equation of (5.39). Consequently, that interval would turn into a one-sided interval when $S = 0$. The presence of $S + 1$ prevents λ_L in (5.39) from being zero even when $S = 0$, to comply with the assumption $\lambda > 0$.

The 95% short and equal-tail CIs are compared visually in Figure 5.22 for the true $\lambda = 2$ using simulations (number of simulations = 100,000). For comparison, we show the coverage probability of the CI computed by the function `PoissonCI`

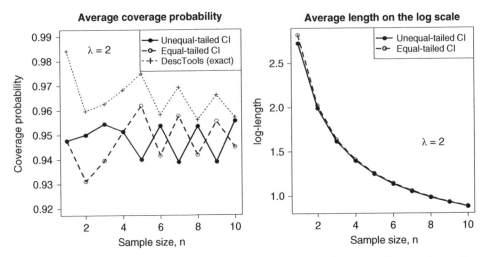

Figure 5.22: *Coverage probability and the average length on the log scale as functions of the sample size for three 95% CIs with the true Poisson rate* $\lambda = 2$.

from the R package `DescTools` with the option `method="exact"`. This method covers the true parameter with a probability slightly higher than the nominal. The equal-tail CI is slightly wider than the short CI, especially for small n.

5.8.2 Two-sided tests and p-value

Now we are working on testing the null hypothesis $H_0 : \lambda = \lambda_0$ versus $H_A : \lambda \neq \lambda_0$. There are two ways to compute quantiles and the respective power function: use (a) the original cdf (5.37) or (b) its continuous counterpart. The former approach is easier but the power function does not go through the point (λ_0, α) due to the discreteness of the Poisson distribution. The latter is more computationally advanced and the power functions go through the point (λ_0, α) but are somewhat artificial because the original data are discrete and therefore do not match simulations, especially for small $n\lambda_0$. Here we follow approach (a).

The quantiles of the equal-tail test are easy to obtain: if q_{L0} and q_{U0} are computed in R as

$$\texttt{qpois(alpha/2, lambda = n * lambda0)}$$

and

$$\texttt{qpois(1 - alpha/2, lambda = n * lambda0)}$$

respectively, we accept the null hypothesis if $q_{L0} < S \leq q_{U0}$. The power of the test is

$$P(\lambda; \lambda_0) = 1 - F(q_{L0}; n\lambda) + F(q_{L0}; n\lambda),$$

where F is the original cdf (5.37).

According to the general equations for the unbiased test (3.1), we want to find continuous quantiles $0 \leq q_L < q_U$ such that

$$C_{2(q_L+1)}(2n\lambda_0) - C_{2(q_U+1)}(2n\lambda_0) - (1-\alpha) = 0,$$

$$c_{2(q_L+1)}(2n\lambda_0) - c_{2(q_U+1)}(2n\lambda_0) = 0. \qquad (5.41)$$

We suggest the R function `nleqslv` from the the package with the same name to solve this system starting from the equal-tail quantiles. After the quantiles are determined testing is simple: the null hypothesis is not rejected if $q_L \leq S \leq q_U$. As for any discrete distribution, the decision about rejection does not depend on the choice of quantiles if they stay within integer bounds. The power function is computed with the same formula as earlier but with q_{L0} and q_{U0} replaced by q_L and q_U computed at the final iteration. The solution of the system (5.41) and computation of the power function are found in the R function `posTest` – the details presented next.

The comparison of the equal- and unequal-tail hypothesis tests for two pairs of λ_0 and n is depicted in Figure 5.23. The theoretical power functions closely match simulation-derived powers following the acceptance rule (number of simulations = 1 million). For small $n\lambda_0$, the theoretical powers do not pass the (λ_0, α) due to the discreteness of (5.37).

Now we discuss the computation of the p-value. As follows from Section 4.1 the p-value of the equal-tail test is computed as

$$p = 2 \times \min(F(S; \lambda_0), 1 - F(S; \lambda_0)),$$

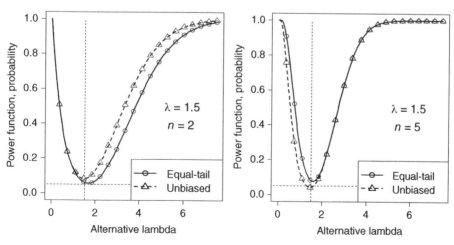

Figure 5.23: *The power functions of the equal-tailed and unbiased Poisson tests for two pairs of λ_0 and n. The respective symbols represent simulations (number of simulations = 1 million). This graph was created with the R function* `posTest(job=1)`.

If the unbiased test is used

$$p = \begin{cases} \frac{\alpha}{F(\theta_L;\lambda_0)} F(S;\lambda_0) & \text{if } F(S;\lambda_0) \leq \frac{1}{\alpha} F(\theta_L;\lambda_0) \\ \frac{\alpha}{\alpha - F(\lambda_L;\theta_0)} (1 - F(S;\lambda_0)) & \text{if } F(S;\lambda_0) > \frac{1}{\alpha} F(\theta_L;\lambda_0) \end{cases} .$$

In both cases, the original cdf (5.37) can be used because S is an integer.

The following example illustrates the computation.

Example 5.13 *Testing abortions.*

The total number of abortions per adult woman younger than 50 years in the state is $\lambda_0 = 1.2$. Five ($n = 5$) randomly asked adult women younger than 50 years in a small village had a total of $S = 11$ abortions. Compute the p-value for testing the hypothesis that the rate of abortions in the village is the same as in the state.

We assume that the total number of abortions by a specific woman follows a Poisson distribution with the rate parameter λ. See the function `posTest(job=2)`. The p-value for the `posTest`equal-tail test is twice the minimum of the two numbers

$$\text{F\_left=ppois(S,lambda=n*lambda0)}$$

$$\text{F\_right=ppois(S,lambda=n*lambda0,lower.tail=F)}$$

where `S=11;n=5;lambda0=1.2`. The p-value for the unbiased test uses the formula shown previously. Following are the call and the output.

```
> posTest(job=2)
Equal-tail test p-value: 0.04018
Quantiles of the equal-tail test: 2 11
Quantiles of the unbiased test: 1.359 10.94227
Unbiased test p-value: 0.03077
```

Both tests produce a p-value less than 0.05.

5.8.3   The MCL estimator of the rate parameter

The classic estimator of the Poisson rate $\hat{\lambda} = \overline{Y}$ has the same problem as the binomial probability when $S = 0$. For example, for $\lambda = 0.5$ and $n = 2$ a positive estimate of the rate does not exist in about 37% of samples. On the contrary, the log scale moves the estimator away from zero. According to our general definition, the MCL estimator is the limit of the CI on the log scale when the confidence level goes to zero. To derive the MCL estimator for the rate parameter divide the second equation of (5.39) by $\lambda_U - \lambda_L$ and let $\alpha \to 1$ yield the MCL estimator of the Poisson rate

$$\hat{\lambda}_{MCL} = \overline{Y} + \frac{1}{n}.$$

This estimator is always positive. For a zero outcome, the estimate of λ is $1/n$.

5.9 Meta-analysis model

Previously we built our theory on the assumption that there is a statistic S with a cdf known up to an unknown parameter. Then the pivotal quantity is the cdf itself evaluated at S because it has a uniform distribution on $(0, 1)$. Remember that the pivotal quantity is a function of a statistic and an unknown parameter whose distribution is parameter-free. Thus the method of the inverse cdf for the CI can be viewed as a special case of the pivotal CI (Shao 2003). Here we illustrate the MC-statistics when instead of a statistic whose distribution depends on an unknown parameter, we have a pivotal quantity. We apply the pivotal quantity approach to optimal statistical inference for the heterogeneity variance in the meta-analysis model, the simplest variance components model (Searle et al. 2006; Demidenko 2013).

Estimation of variance components is challenging because unbiased estimators may take negative values with positive probability. The chance to obtain a negative estimate increases with a small sample size and/or true variance. Moreover, an unbiased variance estimator with positive values does not exist. The goal of this section is to develop the MC estimator of the heterogeneity variance in the meta-analysis model on the log scale as outlined in Section 2.5.1. The maximum concentration estimator on the log scale (MCL estimator) is compared with other popular methods such as the method of moments (MM) and maximum likelihood estimator (MLE).

The meta-analysis model is the simplest variance components model and takes the form

$$Y_i \sim \mathcal{N}(\mu, \tau^2 + \sigma_i^2), \ i = 1, 2, ..., n,$$

where Y_i are independent observations, μ and $\tau^2 \geq 0$ are unknown parameters subject to estimation, and $\sigma_1^2, \sigma_2^2, ..., \sigma_n^2$ are positive known constants. This model is widely used in biostatistics to combine heterogeneous studies for efficient estimation of the treatment effect, μ, where the study-specific variances, σ_i^2 are assumed known.

The point of our concern is the estimation of the variance component, sometimes referred to as the heterogeneity variance τ^2. Two estimators are currently used: the unbiased MM estimator

$$\widehat{\tau}_{MM}^2 = \frac{1}{n-1} \sum_{i=1}^n (Y_i - \overline{Y})^2 - \frac{1}{n} \sum_{i=1}^n \sigma_i^2$$

and the ML estimator, $\widehat{\tau}_{ML}^2$, as the minimizer of the negative log-likelihood function, which up to a constant term, takes the form

$$\sum_{i=1}^n \left[\ln(\tau^2 + \sigma_i^2) + \frac{(Y_i - \widetilde{\mu}(\tau^2))^2}{\tau^2 + \sigma_i^2} \right],$$

where

$$\widetilde{\mu}(\tau^2) = \frac{\sum Y_i/(\tau^2 + \sigma_i^2)}{\sum 1/(\tau^2 + \sigma_i^2)}$$

is the weighted average (Demidenko 2013, Section 5.1).

Both estimators may take negative values as illustrated in Figure 5.24 with $\mu = 1$, the true heterogeneity variance $\tau^2 = 0.5$, the number of studies, $n = 5$; the study-specific variances are randomly generated following the uniform distribution on $(0, 1)$ are held fixed during simulations. From the simulation-derived Gaussian kernel density (number of simulations = 100,000) we conclude that the proportion of ML estimates with negative values is about 16%, and the proportion of negative MM estimates is about 26%.

Our goal is to develop a positive estimator for τ^2 using the pivotal quantity

$$Q(\tau^2) = \sum_{i=1}^{n} \frac{(Y_i - \widetilde{\mu}(\tau^2))^2}{\tau^2 + \sigma_i^2} \sim \chi^2(n-1). \tag{5.42}$$

This quantity is usually referred to as the Q-statistic (DerSimonian and Laird 1986). To make our discussion self-contained we prove (5.42) using the fact that the quadratic form $\mathbf{z'Az} \sim \chi^2(m)$, where $\mathbf{z} \sim \mathcal{N}(\mathbf{0}, \mathbf{I})$ and $\mathbf{A}$ is a symmetric idempotent matrix such that $tr(\mathbf{A}) = m$ (Rao 1973). Our proof follows the proof

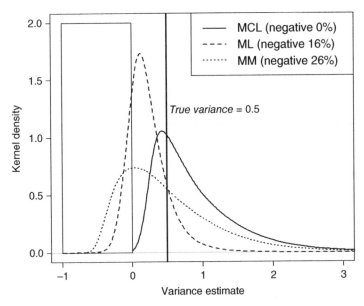

Figure 5.24: *Simulation-derived densities of three estimators of the heterogeneity variance, $\tau^2 = 0.5$ based on $n = 5$ studies. The vertical line depicts the true τ^2, and the grey rectangle depicts negative values. Percent simulations in which estimates were negative are reported in parentheses.*

given by Demidenko (2020, p. 261). The meta-analysis model can be cast into a slope linear regression model $z_i \sim \mathcal{N}(\mu x_i, 1)$, where $z_i = Y_i(\tau^2 + \sigma_i^2)^{-1/2}$ and $x_i = (\tau^2 + \sigma_i^2)^{-1/2}$. The left-hand side of (5.42) is the regression residual sum of squares written as

$$(\mathbf{z} - \tilde{\mu}\mathbf{x})'(\mathbf{I} - \mathbf{x}\mathbf{x}'\|\mathbf{x}\|^{-2})(\mathbf{z} - \tilde{\mu}\mathbf{x}) = \mathbf{z}'\mathbf{A}\mathbf{z},$$

where $\mathbf{z}$ and $\mathbf{x}$ are the $n \times 1$ vectors with components z_i and x_i, respectively, and $\mathbf{A} = \mathbf{I} - \mathbf{x}\mathbf{x}'\|\mathbf{x}\|^{-2}$ is the $n \times n$ idempotent matrix of rank $n - 1$. This proves (5.42).

Derive $\tilde{\mu}(\tau^2)$ for extreme values of the heterogeneity variance. It is elementary to see that

$$\lim_{\tau^2 \to 0} \tilde{\mu}(\tau^2) = \tilde{\mu}_0 = \frac{\sum Y_i/\sigma_i^2}{\sum 1/\sigma_i^2},$$

$$\lim_{\tau^2 \to \infty} \tilde{\mu}(\tau^2) = \lim_{\tau^2 \to \infty} \frac{\sum \tau^2 Y_i/(\tau^2 + \sigma_i^2)}{\sum \tau^2/(\tau^2 + \sigma_i^2)} = \frac{\sum Y_i}{n} = \overline{Y}.$$

Respectively,

$$\lim_{\tau^2 \to 0} Q(\tau^2) = \sum_{i=1}^{n} \frac{(Y_i - \tilde{\mu}_0)^2}{\sigma_i^2} = Q_0,$$

$$\lim_{\tau^2 \to \infty} Q(\tau^2) = 0,$$

where $Q_0 > 0$ with probability 1.

Lemma 5.14 *Function $Q(\tau^2)$ defined by equation (5.42) is a decreasing function of $\tau^2 > 0$ for any Y_i and σ_i^2.*

Proof. It is elementary to check that the derivative of Q can be written as

$$Q' = -\sum \frac{Y_i^2}{(\tau^2 + \sigma_i^2)^2} + \frac{2\left(\sum \frac{Y_i}{\tau^2 + \sigma_i^2}\right)\left(\sum \frac{Y_i}{(\tau^2 + \sigma_i^2)^2}\right)}{\sum \frac{1}{\tau^2 + \sigma_i^2}}$$

$$-\frac{\left(\sum \frac{Y_i}{\tau^2 + \sigma_i^2}\right)^2 \left(\sum \frac{1}{(\tau^2 + \sigma_i^2)^2}\right)}{\left(\sum \frac{1}{\tau^2 + \sigma_i^2}\right)^2}. \tag{5.43}$$

To shorten the notation denote $a_i = Y_i$ and $b_i = 1/(\tau^2 + \sigma_i^2) > 0$. Then the inequality $Q' \leq 0$ is equivalent to

$$S_{22}S_{01}^2 - 2S_{11}S_{12}S_{01} + S_{11}^2 S_{02} \geq 0 \tag{5.44}$$

where

$$S_{22} = \sum a_i^2 b_i^2, \quad S_{01} = \sum b_i, \quad S_{11} = \sum a_i b_i,$$
$$S_{12} = \sum a_i b_i^2, \quad S_{02} = \sum b_i^2.$$

Consider the left-hand side of the inequality (5.44) as a function of $\{a_i, i = 1, ..., n\}$. We aim to prove that

$$\min_{a_1, ... a_n} (S_{22} S_{01}^2 - 2 S_{11} S_{12} S_{01} + S_{11}^2 S_{02}) = 0. \qquad (5.45)$$

Differentiating with respect to a_i, we obtain n conditions

$$2 a_i b_i^2 S_{01}^2 - 2 b_i S_{12} S_{01} - 2 b_i^2 S_{11} S_{01} + 2 b_i S_{11} S_{02} = 0,$$

for $i = 1, ..., n$. The function to minimize is a quadratic function Since $b_i > 0$ these conditions are equivalent to

$$a_i b_i S_{01}^2 - b_i S_{11} S_{01} + (S_{11} S_{02} - S_{12} S_{01}) = 0.$$

Sum over i to obtain

$$S_{11} S_{01}^2 - S_{11} S_{01}^2 + n(S_{11} S_{02} - S_{12} S_{01}) = 0$$

or $S_{11} S_{02} - S_{12} S_{01} = 0$. This yields $a_i b_i S_{01}^2 - b_i S_{11} S_{01} = 0$: that is, the minimum is attained when $a_i = a = $ const. Now we find the minimum

$$S_{22} S_{01}^2 - 2 S_{11} S_{12} S_{01} + S_{11}^2 S_{02}$$
$$= a^2 S_{02} S_{01}^2 - 2 a^2 S_{01} S_{02} S_{01} + a^2 S_{01}^2 S_{02} = 0.$$

It is easy to see that (5.45) implies (5.44) and as such it proves that $Q' \le 0$. $\quad \square$

This result is crucial for establishing the uniqueness of the equal-tail CI.

5.9.1   CI and MCL estimator

The traditional equal-tail exact CI for τ^2 can be derived from the chi-square distribution of the Q-statistic (5.42). Specifically, the lower and upper limits for τ^2 are found by solving $Q(\tau^2) = C_{n-1}^{-1}(1 - \alpha/2)$ and $Q(\tau^2) = C_{n-1}^{-1}(\alpha/2)$, respectively. It may happen that the first solution does not exist when $Q_0 \le C_{n-1}^{-1}(1 - \alpha/2)$. Then we set the lower limit to zero. To solve these equations Newton's algorithm has been applied (Demidenko 2013). The ups2 function solves the equation $Q(\tau^2) = C_{n-1}^{-1}(\psi)$ for any $0 < \psi < 1$. The general call of this function is

```
ups2(y, s2i, psi, maxit = 10, eps = 0.0001)
```

This function is used to compute the lower and upper equal-tail CIs for τ^2. For example, to compute the exact 95% lower limit for τ^2, one issues

`ups2(y, s2i, psi = 0.975)`

and to compute the upper limit one issues

`ups2(y, s2i, psi = 0.025)`

where array y contains the sample $\{Y_i, i = 1, ..., n\}$ and array s2i contains values $\{\sigma_i^2, i = 1, ..., n\}$.

Now we turn our attention to the MC CI on the log scale as discussed in Section 2.5.1. This proposal leads to the minimization of $\ln \tau_U^2 - \ln \tau_L^2$ under the condition

$$C_{n-1}(Q(\tau_L^2)) - C_{n-1}(Q(\tau_U^2)) - \lambda = 0. \tag{5.46}$$

By applying equation (5.2) or employing the Lagrange multiplier technique it is elementary to see that this optimization problem is equivalent to solving the equation

$$\tau_L^2 c_{n-1}(Q(\tau_L^2))Q'(\tau_L^2) - \tau_U^2 c_{n-1}(Q(\tau_U^2))Q'(\tau_U^2) = 0, \tag{5.47}$$

where Q' is given by (5.43). Newton's algorithm requires derivatives of equation (5.47) that may be too complicated. Instead, we apply general algorithms for solving a system of the two nonlinear equations (5.46) and (5.47). The most popular packages in R for solving nonlinear equations are `nleqslv` and `multiroot`. Both are satisfactory, but not without a caveat because the system may have no solution when $Q_0 < C_{n-1}^{-1}(1 - \alpha/2)$; then we set $\tau_L = 0$.

The R function `LOGci` returns a two-dimensional vector as the left-hand sides of (5.46) and (5.47), and function `nleqslv` solves the equations. For example, if `lET` and `uET` are the lower and upper limits of the equal-tail CI and `vari` passes the values of σ_i^2, the call

`nleqslv(x = c(lET, uET), f=LOGci, y = y, s2i = vari,lambda = lambda)$x`

returns the MCL CI for τ^2 as a two-dimensional array as the solution of (5.46) and (5.47).

Figure 5.25 depicts the comparison of the equal-tail and MCL CIs on the range of the heterogeneity variance, τ^2,for $n = 5$ and study-specific variances $1, 2, 3, 4, 5$. Both intervals have the coverage probability close to nominal, $\lambda = 0.95$, although the coverage probability of the MCL CI is more sensitive to small values of Q_0 when the computation of the lower limit becomes numerically unstable. The length of the MCL CI computed on the log scale is uniformly smaller than that of equal-tail CI.

The MCL estimator of heterogeneity variance on the log scale, $\widehat{\tau}^2_{MCL}$, is the limit point of the system (5.46) and (5.47) when $\lambda \to 0$ and is found as the minimizer of the function (`Mtau2` in R)

$$M(\tau^2) = \tau^2 c_{n-1}(Q(\tau^2))Q'(\tau^2).$$

It is elementary to prove that $\widehat{\tau}^2_{MCL}$ is positive because

$$\lim_{\tau \to 0} M(\tau^2) = \lim_{\tau^2 \to \infty} M(\tau^2) = 0$$

and $M'(0) < 0$ with probability 1. Instead of Newton's algorithm as used before, we suggest a general algorithm for the minimization of $M(\tau^2)$ such as `optimize` as a part of base R. The call

```
optimize(Mtau2, interval = c(0,100), y = Y, s2i = vari)$minimum
```

returns the estimate $\widehat{\tau}_{MCL}$, where `interval` specifies a user-defined interval for the minimizer, array `y` contains values of Y_i, and array `s2i` contains values of variances σ_i^2.

The Gaussian kernel density of $\widehat{\tau}^2_{MCL}$ with $n = 5$ and $\tau^2 = 0.5$ is depicted in Figure 5.24. The MCL estimates are always positive with the maximum density around the true heterogeneity variance of 0.5.

Figure 5.26 demonstrates the optimal property of the MCL estimator using simulations – the values have the highest concentration around the true heterogeneity variance than the ML and MM estimates under the same scenario as in

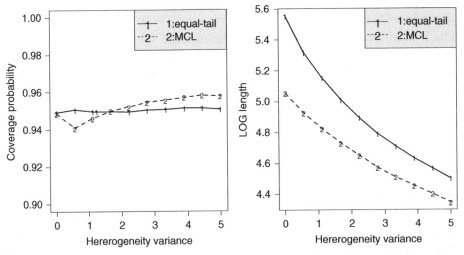

Figure 5.25: *Comparison of the equal-tail and MCL CIs for heterogeneity variance, τ^2, with $\sigma_i^2 = i$, $i = 1, 2, ..., n = 5$ using simulations (number of simulations = 100,000).*

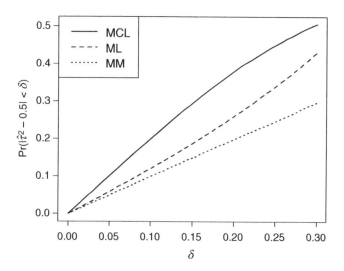

Figure 5.26: *Concentration probabilities around the true* $\tau^2 = 0.5$ *against* δ. *The MC estimator on the log scale (MCL estimator) has the maximum concentration probability.*

Figure 5.24. On the y-axis, we plot the proportion of the three estimates that fall within the interval $0.5 \pm \delta$ where δ varies from 0 to 0.3. This proportion is higher for the MCL estimator on the entire range of δ. Simply put, $\widehat{\tau}^2_{MCL}$ is closer to the true heterogeneity variance compared to ML and MM estimators.

5.10 M-statistics for the correlation coefficient

The goal of this section is to apply M-statistics to the Pearson correlation coefficient

$$r = \frac{\sum_{i=1}^{n}(X_i - \overline{X})(Y_i - \overline{Y})}{\sqrt{\sum_{i=1}^{n}(X_i - \overline{X})^2 \sum_{i=1}^{n}(Y_i - \overline{Y})^2}},$$

where (X_i, Y_i) are iid observations from a bivariate normal distribution with the population correlation coefficient ρ. The exact statistical inference dates back to the fundamental work by David (1938) using the equal-tail approach. Although several approximations to the CI and hypothesis testing have been suggested over the years, no other exact statistical inference is known to date – see Nie et al. (2011) and Hu et al. (2020) for review. Ironically, the work of David (1938) published in the form of tables, is forgotten, despite today's unmatched computer power. Although several attempts have been made to implement David's tables (Anderson 2003), the exact statistical inference for the correlation coefficient, especially important for small and extremely small sample sizes, remains underdeveloped. Even modern statistical packages do not offer exact power functions and p-values for the nonzero null hypothesis.

The goal of this section is to revive the exact statistical inference for the correlation coefficient and develop new exact optimal statistical inference including CI, hypothesis testing, and the respective p-values. No less important are numerical algorithms for the double-sided exact statistical tests and CIs – the theoretical developments are often checked via simulations. The novel methods are especially valuable for extremely small sample sizes such as $n = 6$. When possible, we apply Newton's algorithm because a successful starting value based on the Fisher Z-transformation is readily available. However, when Newton's iterations require high-order derivatives, or in the case of convergence issues, we apply general built-in R functions such as `uniroot` or `optimize`.

The density of the correlation coefficient was derived by Fisher (1915). To improve convergence, Hotelling (1953) suggested expressing the density through the Gaussian hypergeometric function (see also Anderson 2003, p. 125) as

$$
f(x; \rho) = \frac{(n-2)\Gamma(n-1)(1-\rho^2)^{\frac{n-1}{2}}(1-x^2)^{\frac{n-4}{2}}}{\sqrt{2\pi}\Gamma(n-1/2)(1-\rho x)^{\frac{2n-3}{2}}} G\left(\frac{1}{2}, \frac{1}{2}, n - \frac{1}{2}; \frac{\rho x + 1}{2}\right),
\tag{5.48}
$$

where $G(a, b, c; z)$ stands for the Gaussian hypergeometric function. Computation of the hypergeometric function is readily available in modern statistical packages – we use the package `hypergeo` in R, although numerical integration via `integrate` in R based on the integral representation of G given shortly is another option. The package `gsl` is an alternative to `hypergeo`. The cdf of r is defined via the integral over the density as

$$
F(x; \rho) = \int_{-1}^{x} f(u; \rho) du
\tag{5.49}
$$

numerically evaluated by the R function `integrate`. Although several attempts have been made to approximate the cdf of r we found that the straightforward numerical integration of the density (5.48) yields the most satisfactory results.

5.10.1   MC and MO estimators

As follows from Section 2.2, the MC estimator, r_{MC}, is defined as the value ρ for which the slope of F as a function of ρ, given the observed r: that is, $F'(r; \rho)$ reaches its minimum, where r is the observed Pearson correlation coefficient:

$$
r_{MC} = \arg\min_{\rho} \int_{-1}^{r} \frac{\partial f(x; \rho)}{\partial \rho} dx,
\tag{5.50}
$$

where f is given by (5.48). To compute the derivative of the density under the integral we use the following integral representation:

$$G(a, b, c; z) = \frac{\Gamma(c)}{\Gamma(b)\Gamma(c - b)} \int_0^1 t^{b-1}(1 - t)^{c-b-1}(1 - tz)^{-a}dt$$

along with its derivative with respect to z given by

$$\frac{\partial}{\partial z}G(a, b, c; z) = \frac{ab}{c}G(a + 1, b + 1, c + 1; z), \quad k = 1, 2, \ldots \qquad (5.51)$$

Since the derivative of f with respect to ρ is easier to express after the log transformation we use the formula

$$\frac{\partial F(r; \rho)}{\partial \rho} = \int_{-1}^r \frac{\partial f(x; \rho)}{\partial \rho}dx = \int_{-1}^r \frac{\partial \ln f(x; \rho)}{\partial \rho}f(x; \rho)dx. \qquad (5.52)$$

Define

$$H_k = G\left(\frac{1}{2} + k, \frac{1}{2} + k, n - \frac{1}{2} + k; \frac{x\rho + 1}{2}\right)$$

for $k = 0, 1, 2, 3$. Then the recurrence formula (5.51) implies

$$\frac{\partial H_k}{\partial \rho} = \frac{x}{4}\frac{(2k + 1)^2}{2n + 2k - 1}H_{k+1}.$$

Finally,

$$\frac{\partial \ln f(x; \rho)}{\partial \rho} = -\frac{\rho(n - 1)}{1 - \rho^2} + \frac{x(2n - 3)}{2(1 - \rho x)} + \frac{x}{4(2n - 1)}\frac{H_1}{H_0}. \qquad (5.53)$$

To compute r_{MC}, we use the R function `optimize` with the integral in (5.50) evaluated using function `integrate`.

Figure 5.27 illustrates the MC estimate as the inflection point on the curve $F(r; \rho)$ – the point of the steepest slope indicated by a circle at seven observed values of the correlation coefficient, r, with $n = 10$. This figure was created with the function `moder_MC`.

The fact that r is biased has been known for hundreds of years. Olkin and Pratt (1958) developed an unbiased estimator expressed through the G function as

$$r_{OP} = rG(1/2, 1/2, (n - 1)/2, 1 - r^2).$$

The MO estimator, r_{MO}, maximizes the pdf (5.48) given the observed value r (the R function `optimize` is used again). Figure 5.28 illustrates the difference between the Olkin-Pratt, MC, and MO estimators as the deviation from the observed r when $n = 10$. We highlight several points: (a) the difference is symmetric around 0, (b) the difference reduces with n, (c) the maximum absolute difference occurs around $r = \pm 0.6$, (d) the Olkin-Pratt estimator deviates less from the observed correlation coefficient than MC estimator, and (e) the MO estimator deviates from r in the opposite direction than the MC estimator.

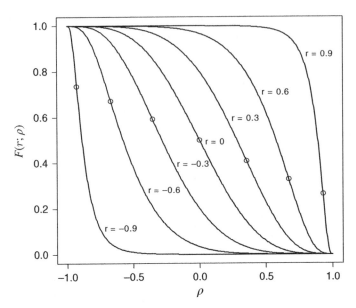

Figure 5.27: *The MC estimate of ρ shown by a circle as the inflection point on the curve F(r; ρ) for seven values of r with the sample size n = 10. This graph was created with the function* moder_MC.

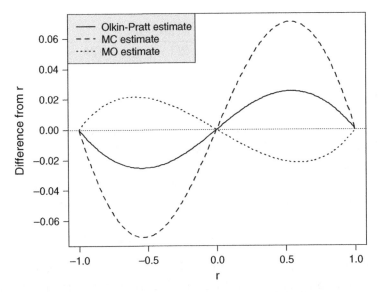

Figure 5.28: *Comparison of the Olkin-Pratt, MC, and MO estimates of the correlation coefficient as the difference from the observed r with the sample size n = 10. This graph was created with the function* moder_MO.

5.10.2   Equal-tail and unbiased tests

Now we turn our attention to double-sided hypothesis testing

$$H_0 : \rho = \rho_0 \text{ against } H_0 : \rho \neq \rho_0. \tag{5.54}$$

As we mentioned earlier, the existing statistical inference for the correlation coefficient relies on Fisher's Z-transformation defined as

$$Z(x) = \frac{1}{2} \ln \frac{1+x}{1-x}, -1 < x < 1.$$

Fisher (1915) proved that

$$Z(r) \simeq \mathcal{N}(Z(\rho), (n-3)^{-1})$$

for large n. Surprisingly, this approximation works well for small n as well, although is not exact. For example, using this approximation, the null hypothesis is rejected if the $100\lambda\%$ CI on the Z-scale, $Z(r) \pm \Phi^{-1}((1+\lambda)/2)/\sqrt{n-3}$ does not contain $Z(\rho_0)$ with the p-value given by

$$p = 2\Phi(-|Z(r) - Z(\rho_0)|\sqrt{n-3}). \tag{5.55}$$

The back transform of the CI on the Z-scale produces an approximate CI for ρ – for details, see Demidenko (2020, p. 580). In the language of cdf this approximation yields

$$F(x; \rho) \simeq \Phi((Z(x) - Z(\rho))\sqrt{n-3}), \tag{5.56}$$

to be referred to as the Fisher Z-transformation cdf approximation. For the Fisher approximation, the equal-tail quantiles q_L and q_U are computed as

$$q = Z^{-1}\left(Z(\rho_0) + (n-3)^{-1/2}\Phi^{-1}(\eta)\right),$$

where

$$Z^{-1}(z) = \frac{e^{2z} - 1}{e^{2z} + 1}$$

is the inverse Z-transform with $\eta = \alpha/2$ and $\eta = 1 - \alpha/2$ that produce q_L and q_U, respectively.

The exact equal-tail test requires finding the $(\alpha/2)$ and $(1 - \alpha/2)$ quantiles of the cdf (5.49) evaluated at $\rho = \rho_0$ defined as

$$F(q_L; \rho_0) = \alpha/2, \ F(q_U; \rho_0) = 1 - \alpha/2.$$

These equations are effectively solved by Newton's algorithm

$$q_{s+1} = q_s - \frac{F(q_s; \rho_0) - \eta}{f(q_s; \rho_0)}, \quad s = 0, 1, \dots$$

where $\eta = \alpha/2$ and $\eta = 1 - \alpha/2$ produce q_L and q_U, respectively, starting from Fisher's quantiles.

Now we aim to construct an exact unbiased two-sided test for the null hypothesis $H_0 : \rho = \rho_0$ versus the alternative $H_A : \rho \neq \rho_0$ having the exact size α.

As follows from Section 2, the unbiased test reduces to finding quantiles $-1 < q_L < q_U < 1$ such that

$$F(q_U; \rho_0) - F(q_L; \rho_0) = \lambda, \ F'(q_L; \rho_0) - F'(q_U; \rho_0) = 0, \tag{5.57}$$

where F' denotes the derivative of the cdf (5.49) with respect to ρ using formula (5.52). The first equation guarantees size α, and the second equation guarantees the unbiasedness. The null hypothesis is rejected if r falls outside the interval (q_L, q_U). The power function of the unbiased test is given by (5.58).

Now we turn our attention to solving the system (5.57). Again, Newton's algorithm applies with the adjustment vector for $(q_L, q_U)'$ given by

$$\begin{bmatrix} -f(q_L; \rho_0) & f(q_U; \rho_0) \\ f'(q_L; \rho_0) & -f'(q_U; \rho_0) \end{bmatrix}^{-1} \begin{bmatrix} F(q_U; \rho_0) - F(q_L; \rho_0) - \lambda \\ F'(q_L; \rho_0) - F'(q_U; \rho_0) \end{bmatrix},$$

where f' is the derivative of the density (5.48), starting from the quantiles computed from the Fisher Z-transformation or equal-tail.

The R function test_r prints out the simulation-based type I errors along with the theoretical error computed via the cdf (5.49). For extreme values of ρ_0 and small n, the noncentral t-approximation yields results that are not as good as those for the Fisher Z-approximation. This code contains functions for computing the cdf and the pdf of r along with their derivatives – more details are provided in the next section.

5.10.3   Power function and p-value

Once quantiles q_L and q_U are determined, the rejection rule is the same: the null hypothesis $H_0 : \rho = \rho_0$ is rejected if ρ_0 is outside the interval (q_L, q_U). As follows from Section 2.3, the power function of the test is given by

$$P(\rho; \rho_0) = 1 + F(q_L; \rho) - F(q_U; \rho). \tag{5.58}$$

Function test_r prints out the simulation-based and theoretical power function (5.58). A typical call with the output is listed next. For extreme values of the sample size and the alternatives, the power values between exact and Fisher's Z-test may reach up to a 20% difference while the difference between the unbiased and equal-tail test is modest.

```
> test_r(n=10,ro0=.25,ro.alt=.5,r=.1,alpha=0.05,maxit=10,ss=3,nSim=1000000)
Time difference of 5.030669 mins
                Sim type I  Theor type I Sim power Theor power p-value power
Exact equal-tail  0.050197  0.05000000  0.123089   0.1227653        0.123089
Fisher approx     0.051211  0.05091585  0.132844   0.1326054        0.132844
Unbiased exact    0.050224  0.05000000  0.123958   0.1236705        0.123958
```

Figure 5.29 depicts the power functions of three tests of size $\alpha = 0.05$ computed by the same formula (5.58) but with different q_L and q_U with $n = 10$, $\rho_0 = 0.8$, and $\alpha = 0.05$. We make several comments: (a) only for the unbiased test does the power function touch the horizontal line $P = 0.05$ and do the other two curves intersect the line. (b) All power curves pass the point $(0.8, 0.05)$ except the Fisher Z-test but it is very close. (c) The equal-tail and unbiased tests have similar performance.

Now we discuss the computation of the p-value for the three tests. The formula for the p-value of the Fisher Z-test is given by (5.55). As follows from Section 2.3 the exact p-value for the equal-tail and unbiased tests is computed as

$$p = \begin{cases} \frac{\alpha}{F(q_L;\rho_0)}F(r;\rho_0) & \text{if } F(r;\rho_0) \le \frac{1}{\alpha}F(q_L;\rho_0) \\ \frac{\alpha}{\alpha-F(q_L;\rho_0)}(1 - F(r;\rho_0)) & \text{if } F(r;\rho_0) > \frac{1}{\alpha}F(q_L;\rho_0) \end{cases}.$$

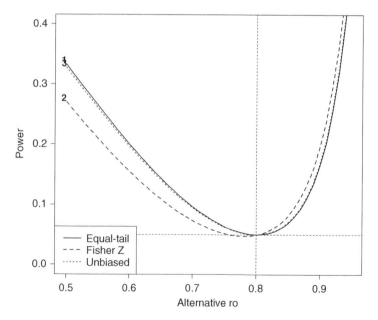

Figure 5.29: *The power function for three tests with* $n = 10$, $\rho_0 = 0.8$, *and* $\alpha = 0.05$. *This graph was created with the R function* **test_rpow**.

Since for the equal-tail $F(q_L; \rho_0) = \alpha/2$, it simplifies to

$$p = 2 \times \min(F(r; \rho_0), 1 - F(r; \rho_0)).$$

We refer the reader to the last column in the output of the call to the test_r function where the power is computed as the proportion of simulations for which $p < \alpha$. The match between the simulation-based p-value and theoretical power is very good.

5.10.4 Confidence intervals

The CIs on using Fisher Z-transformation are well known and can be computed through the Z-inverse as

$$Z^{-1}(Z(r) \pm \Phi^{-1}((1 + \lambda)/2)/\sqrt{n - 3}), \tag{5.59}$$

where $+$ produces the upper limit (ρ_L) and $-$ produces the lower limit (ρ_U). The exact equal-tail CI for ρ was developed by David (1938) and presented in the form of tables. We obtain the exact confidence limits ρ_L and ρ_U by solving the equation

$$\int_{-1}^{r} f(u; \rho)du = \eta,$$

where the lower limit (ρ_L) is derived by letting $\eta = 1 - \alpha/2$ and the upper limit (ρ_U) is derived by letting $\eta = \alpha/2$. The limits are computed using Newton's algorithm:

$$\rho_{s+1} = \rho_s - \frac{F(r; \rho_s) - \eta}{F'(r; \rho_s)}, \quad s = 0, 1, ..., \tag{5.60}$$

where the formulas for the integrands were given previously with the iterations starting from the Fisher Z-transformation limits (5.59). These limits are readily translated into hypothesis testing using the dual principle: the null hypothesis $H_0 : \rho = \rho_0$ is rejected if the interval (ρ_L, ρ_U) does not cover ρ_0.

The lower and upper limits of the unbiased CI (ρ_L, ρ_U) are found by solving the system of equations

$$F(r; \rho_L) - F(r; \rho_U) - \lambda = 0, \quad \ln f(r; \rho_L) - \ln f(r; \rho_U) = 0.$$

This system is solved iteratively by Newton's algorithm with the adjustment vector for (ρ_L, ρ_U) given by

$$\begin{bmatrix} F'(r; \rho_L) & -F'(r; \rho_U) \\ \frac{\partial \ln f(r; \rho_L)}{\partial \rho} & -\frac{\partial \ln f(r; \rho_U)}{\partial \rho} \end{bmatrix}^{-1} \begin{bmatrix} F(r; \rho_L) - F(r; \rho_U) - \lambda \\ \ln f(r; \rho_L) - \ln f(r; \rho_U) \end{bmatrix},$$

where F' means differentiation with respect to the second argument/parameter and computed by formula (5.52) with the derivative of the log density given by (5.53), starting from the equal-tail CI limits.

The short CI for ρ is computed by solving the system of equations

$$F(r; \rho_L) - F(r; \rho_U) - \lambda = 0, \quad F'(r; \rho_L) - F'(r; \rho_U) = 0.$$

Newton's adjustment vector takes the form

$$\begin{bmatrix} F'(r; \rho_L) & -F'(r; \rho_U) \\ F''(r; \rho_L) & -F''(r; \rho_U) \end{bmatrix}^{-1} \begin{bmatrix} F(r; \rho_L) - F(r; \rho_U) - \lambda \\ F'(r; \rho_L) - F'(r; \rho_U) \end{bmatrix}.$$

To compute the second derivative of F, we use the recurrence formula (5.51), which yields

$$\frac{\partial^2 \ln f}{\partial \rho^2} = -\frac{(1 + \rho^2)(n - 1)}{(1 - \rho^2)^2} + \frac{r^2(2n - 3)}{2(1 - \rho r)^2} + \frac{r^2}{16(2n - 1)}$$
$$\times \left[\frac{9}{2n + 1} \frac{H_2}{H_0} - \frac{1}{2n - 1} \frac{H_1^2}{H_0^2} \right].$$

From (5.52), it follows that

$$F''(r; \rho) = \int_{-1}^{r} \left(\frac{\partial^2 \ln f}{\partial \rho^2} + \left(\frac{\partial \ln f}{\partial \rho} \right)^2 \right) f(r; \rho) du.$$

In Section 5.10.1, the MC estimate r_{MC} was computed using the optimize function. The previous formula for F'' suggests an alternative way to compute the MC estimate $r_{MC} = q$ by solving the equation $F''(r; q) = 0$ using the function uniroot. Unfortunately, both functions require specification of the interval which contains the solution but a tempting choice $(-1, 1)$ does not work. Instead, one may suggest seeking the solution, say, contained in $(-0.999, 0.999)$.

We illustrate the short CI by letting $\lambda \to 0$. By Definition 2.5 the short CI converges to the MC estimate $\widehat{\rho}_{MC}$ depicted in Figure 5.30. The MC estimate shown by a circle at $\lambda = 0$ is computed as discussed in Section 5.10.1 using the function optimize and cross depicts the MC estimate by using uniroot with two pairs of the observed r and the sample size.

Figure 5.31 depicts four CIs for ρ as a function of the observed r with extremely low sample size $n = 6$; when n is moderately large, say, $n = 20$ the difference between intervals is negligible.

Below is the output of the function conf_r with the simulation-derived coverage probabilities for four methods.

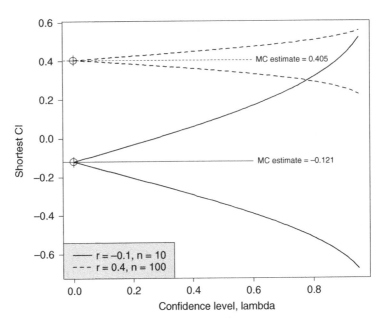

Figure 5.30: *The MC estimate as the limit of the optimized CI when* $\lambda \to 0$ *for the observed values* $r = -0.1$ *and* $r = 0.4$ *with respective sample sizes.*

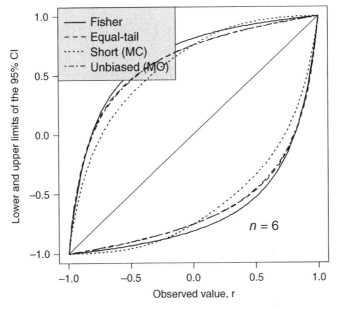

Figure 5.31: *Four 95% CIs as a function of the observed correlation coefficient* r *for* $n = 6$. *This graph was created by calling* `moder(job=9)`.

```
> conf_r(n=10,ro0=.25,alpha=0.05,maxit=5,ss=5,nSim=10000)
Time difference of 18.7254 mins
                Coverage probability
Fisher approx           0.9497
Exact equal-tail        0.9501
Unbiased exact          0.9492
Short exact             0.9193
```

The short CI suffers from integration instability due to the spike-shaped form of the density (5.48) when r is close to ± 1. This results in a slight drop of the coverage probability when n is extremely small. On the other hand, the equal-tail and unbiased CI are unaffected by the spike-shape form and produce a coverage probability practically equal to nominal.

5.11 The square multiple correlation coefficient

The square multiple correlation coefficient, or multiple correlation coefficient (MCC) for short, is one of the central statistical parameters and yet the literature on its statistical inference, including exact CI and unbiased statistical hypothesis testing, is sparse and incomplete. So far only approximate equal-tail and asymptotic inference have been developed. The goal of this section is to fill the gap in the optimal exact statistical inference for the MCC. In particular, the derived MC estimator of the MCC is always positive, unlike its unbiased counterpart.

The MCC is a peculiar example of statistical inference for which the cdf is not equal to 1 at $\rho^2 = 0$ for any given observed MCC. Therefore, condition 4 from Section 2.1 does not hold, although some easy fixes apply as outlined in Section 4.4.

It is assumed that n multidimensional observations $(y_i, \mathbf{x}_i)$ are independently drawn from a multivariate normal distribution $(y, \mathbf{x})$, where y is a scalar and $\mathbf{x}$ is a p-dimensional vector. With the notation $var(y) = \sigma^2$, $cov(y, \mathbf{x}) = \boldsymbol{\omega}$, $cov(\mathbf{x}) = \boldsymbol{\Omega}$, the population MCC is defined as $\rho^2 = \sigma^{-2} \boldsymbol{\omega}' \boldsymbol{\Omega}^{-1} \boldsymbol{\omega}$. Its sample counterpart is given by

$$r^2 = \frac{\mathbf{y}' \mathbf{X} (\mathbf{X}' \mathbf{X})^{-1} \mathbf{X} \mathbf{y}}{\|\mathbf{y}\|^2}, \tag{5.61}$$

where $\mathbf{y}^{n \times 1}$ and $\mathbf{X}^{n \times p}$ are the centered vector and matrix consisting of n observations, respectively.

The density of r^2 was derived by Fisher (1928):

$$f(x; \rho^2) = \frac{\Gamma\left(\frac{n-1}{2}\right)}{\Gamma\left(\frac{n-p-1}{2}\right)\Gamma\left(\frac{p}{2}\right)} (1 - \rho^2)^{\frac{n-1}{2}} x^{\frac{p-2}{2}} (1-x)^{\frac{n-p-3}{2}} G\left(\frac{n-1}{2}, \frac{n-1}{2}, \frac{p}{2}; \rho^2 x\right), \tag{5.62}$$

where G denotes the Gaussian hypergeometric function defined in Section 5.10.1. The R package hypergeo is used to compute G. Similarly to the correlation coefficient, the distribution of r^2 depends not on other parameters of the multivariate normal distribution but on ρ^2 itself – a rather remarkable result. It is well known that r^2 has a positive bias. Olkin and Pratt (1958) developed an unbiased estimator

$$r_{OP}^2 = 1 - \frac{n-3}{n-p-1}(1-r^2)G(1, 1, (n-p+1)/2; 1-r^2).$$

Unfortunately, it may take negative values. For example, for small r^2, we have $r_{OP}^2 \simeq -p/(n-p-2) < 0$.

Several authors, including (Lee 1971, 1972), developed approximations to the cdf of r^2, but they all have moderate quality for extremely small n. For example, for $n = 5$ and $p = 2$, the cdf based on the noncentral F-distribution underestimates the true cdf by 0.12. Instead, we compute the cdf of r^2 by numerical integration of the density (5.62) because we need the derivatives with respect to ρ^2. We have to note that relatively poor approximations and difficulties of exact statistical inference for MCC can be partly explained by a somewhat aberrant behavior of the density (5.62), especially for small ρ^2 and a small number of predictors. For instance, for $p = 1$ (perhaps the most important case) and a small sample size, such as $n < 10$, the density has a local minimum and goes to infinity when $x \to 0$. A relevant discussion is found in Section 5.12.1. These difficulties translate into numerical instability when computing the cdf and its derivatives with respect to ρ^2.

The organization of this section is as follows. First, we derive the unbiased double-sided statistical test and compute the p-value. Second, we discuss obtaining the exact confidence level for one- and two-sided CIs. Finally, we derive the MCL estimator and compare it with the unbiased Olkin-Pratt estimator. Needless to say, more work has to be done on the development of reliable numerical algorithms.

5.11.1 Unbiased statistical test

The goal of this section is to develop an unbiased statistical test for the null hypothesis $H_0 : \rho^2 = \rho_0^2$ versus $H_A : \rho^2 \neq \rho_0^2$. Previous authors derived the equal-tail test using an approximation to the cdf based on the work of Lee (1971, 1972) implemented in the R package MBESS. The respective test and the associated power function are used to determine the sample size required to achieve the desired power.

As follows from Section 3.1, to derive the unbiased test of size $\alpha = 1 - \lambda$, we solve the system of equations

$$F(q_U; \rho_0^2) - F(q_L; \rho_0^2) - \lambda = 0, \quad F'(q_L; \rho_0^2) - F'(q_U; \rho_0^2) = 0 \quad (5.63)$$

for quantiles $0 \leq q_L < q_U$, where F is the cdf of the MSCC obtained through integration, and F' is the derivative with respect to ρ^2:

$$F'(x; \rho^2) = \int_0^x f'(s, \rho^2) ds. \tag{5.64}$$

Once the quantiles are determined, the null hypothesis is rejected if the observed r^2 falls outside the interval (q_L, q_U). Equivalently, the hypothesis is accepted if $F(q_L; \rho_0^2) < F(r^2; \rho_0^2) < F(q_U; \rho_0^2)$. The system (5.63) can be solved by Newton's algorithm with the adjustment vector

$$\begin{bmatrix} f(q_L; \rho_0^2) & -f(q_U; \rho_0^2) \\ -f'(q_L; \rho_0^2) & f'(q_U; \rho_0^2) \end{bmatrix}^{-1} \begin{bmatrix} F(q_U; \rho_0^2) - F(q_L; \rho_0^2) - \lambda \\ F'(q_L; \rho_0^2) - F'(q_U; \rho_0^2) \end{bmatrix}$$

starting from the equal-tail quantiles (see the following). Typically, it requires three or four iterations to converge. The required derivatives up to the third order are provided next.

Proposition 5.15 *Derivatives of the cdf of the square correlation coefficient.*

The derivatives are computed in the manner of the correlation coefficient from the previous section. Denote

$$H_k = G\left(\frac{n-1}{2} + k, \frac{n-1}{2} + k, \frac{p}{2} + k; r^2\rho^2\right), \quad k = 0, 1, 2, 3,$$

where G denotes the Gaussian hypergeometric function. The following identity is handy for computing the derivatives of the pdf with respect to ρ^2:

$$\frac{\partial H_k}{\partial \rho^2} = \frac{r^2}{2} \frac{(n+2k-1)^2}{p+2k} H_{k+1}.$$

Applying this identity, the derivatives of the first, second, and the third order of the log density function with respect to ρ^2 are

$$\frac{\partial \ln f}{\partial \rho^2} = -\frac{n-1}{2(1-\rho^2)} + \frac{r^2(n-1)^2}{2p} \frac{H_1}{H_0},$$

$$\frac{\partial^2 \ln f}{\partial \rho^4} = -\frac{n-1}{2(1-\rho^2)^2} + \frac{r^4(n-1)^2}{4p}\left[-\frac{(n-1)^2 H_1^2}{p H_0^2} + \frac{(n+1)^2 H_2}{(p+2)H_0}\right],$$

$$\frac{\partial^3 \ln f}{\partial \rho^6} = -\frac{n-1}{(1-\rho^2)^3} + \frac{r^4(n-1)^2}{4p}\left[-\frac{(n-1)^2 A}{p} + \frac{(n+1)^2 B}{p+2}\right],$$

where
$$A = \frac{\partial}{\partial \rho^2} \frac{H_1^2}{H_0^2} = -r^2 \frac{(n-1)^2}{p} \frac{H_1^3}{H_0^3} + r^2 \frac{(n+1)^2}{p+2} \frac{H_1 H_2}{H_0^2},$$

$$B = \frac{\partial}{\partial \rho^2} \frac{H_2}{H_0} = -\frac{r^2}{2} \frac{(n-1)^2}{p} \frac{H_2 H_1}{H_0^2} + \frac{r^2}{2} \frac{(n+3)^2}{p+4} \frac{H_3}{H_0}.$$

Finally, the derivatives of the cdf $F(r^2; \rho^2)$ take the form

$$\frac{\partial F}{\partial \rho^2} = \int_0^{r^2} \frac{\partial \ln f}{\partial \rho^2} f(u; \rho^2) du,$$

$$\frac{\partial^2 F}{\partial \rho^4} = \int_0^{r^2} \left(\frac{\partial^2 \ln f}{\partial \rho^4} + \left(\frac{\partial \ln f}{\partial \rho^2} \right)^2 \right) f(u; \rho^2) du,$$

$$\frac{\partial^3 F}{\partial \rho^6} = \int_0^{r^2} \left(\frac{\partial^3 \ln f}{\partial \rho^6} + 3 \frac{\partial \ln f}{\partial \rho^2} \frac{\partial^2 \ln f}{\partial \rho^4} + \left(\frac{\partial \ln f}{\partial \rho^2} \right)^3 \right) f(u; \rho^2) du.$$

$\square$

The power function takes the familiar form

$$P(\rho^2; \rho_0^2) = 1 + F(q_L; \rho^2) - F(q_U; \rho^2), \tag{5.65}$$

where the quantiles q_L and q_U, as functions of ρ_0^2, are found by solving the system (5.63).

The quantiles for the equal-tail test are found from solutions $F(q_U; \rho_0^2) = (1 + \lambda)/2$ and $F(q_L; \rho_0^2) = (1 - \lambda)/2$. Both equations can be effectively solved by Newton's algorithm

$$q_{s+1} = q_s - \frac{F(q_s; \rho_0^2) - \eta}{f(q_s; \rho_0^2)}, \quad s = 0, 1, ...,$$

where $\eta = (1 + \lambda)/2$ for q_U and $\eta = (1 - \lambda)/2$ for q_L. We found that a good starting point for q_L is ρ_0^2 and for q_U is $(1 + \rho_0^2)/2$.

Figure 5.32 shows power functions for $n = 30$ and $p = 7$ with two null values $\rho^2 = 0.6$ and $\rho^2 = 0.3$ (the MBESS package, which uses a cdf approximation, failed in these scenarios). The power of the equal-tail test is very close to the unbiased, so it is not seen. Note that the derivative of the power function at ρ_0^2 vanishes because of the unbiasedness. An important observation is that the power does not reach 1 when the null value approaches 0. This graph was created with the R function unbtestr2 (the package hypergeo must be installed beforehand). This function computes q_L and q_U from the system (5.63) and carries out simulations to confirm the analytical power (5.65) via the rejection rule: the hypothesis is rejected if r^2 falls outside the interval (q_L, q_U).

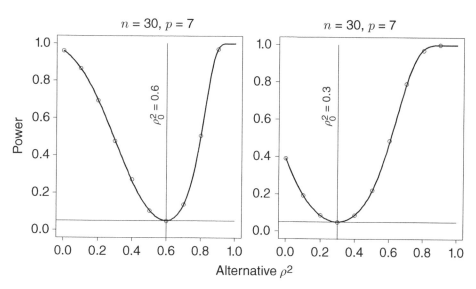

Figure 5.32: *The power function of the unbiased test for two null values, $\rho_0^2 = 0.6$ and $\rho_0^2 = 0.3$. Circles indicate the power values obtained from simulations using the rejection rule. Note that the power with $\rho_0^2 = 0.3$ does not reach 1 when the alternative ρ^2 approaches 0.*

5.11.2 Computation of p-value

Derivation of the p-value in the framework of M-statistics was discussed in Section 4. According to Definition 4.1, the p-value for the MSCC is computed as follows:

$$
p = \begin{cases} \frac{\alpha}{F(q_L;\rho_0^2)} F(r^2;\rho_0^2) & \text{if } F(r^2;\rho_0^2) \leq \frac{1}{\alpha}F(q_L;\rho_0^2) \\ \frac{\alpha}{\alpha - F(q_L;\rho_0^2)}(1 - F(r^2;\rho_0^2)) & \text{if } F(r^2;\rho_0^2) > \frac{1}{\alpha}F(q_L;\rho_0^2) \end{cases},
$$

where q_L and q_U are computed from either the unbiased (5.63) or the equal-tail test ($\alpha = 1 - \lambda$). As follows from Section 4, since $F(q_L;\rho_0^2) = \alpha/2$, the p-value is double the tail probability. For example, in a scenario from Figure 5.32 with the true $\rho_0^2 = 0.6$ and observed $r^2 = 0.4$. Since for the first case, $\rho_0^2 = 0.6$, we have $q_L = 0.4545316$ and $q_U = 0.8488540$ and as follows from the run of `unbtestr2` we compute $p = 0.01895613$. The output of this program is as follows.

```
unbtestr2()
Observed r2=0.4, r20=0.6, p-value=0.018956
qL=0.454532, qU=0.848854
Observed r2=0.4, r20=0.3, p-value=0.664389
qL=0.199193, qU=0.70273
Time difference of 19.25814 secs
```

5.11.3　Confidence intervals

Previous authors proposed one- and two-sided CIs using cdf approximations, including Kraemer (1973), Steiger and Fouladi (1992), Algina and Olejnik (2000), Mendoza et al. (2000). According to the method of inverse cdf, the upper-sided CI $(r_L^2, 1)$ requires a solution of the equation

$$F(r^2; r_L^2) = \lambda. \tag{5.66}$$

It is expected that the interval covers the true ρ^2 with probability λ. However, equation (5.66) may have no solution if $F(r^2; 0) < \lambda$. This undesirable inequality happens particularly often for small r^2 and small df. Figure 5.33 illustrates the case. Function $F(r^2, \rho^2)$ is plotted for the same r^2 and $p = 6$ but different n. While the upper-sided 95% CI exists for $n = 30$, it does not for $n = 12$ (the circle indicates the probability of the beta distribution). However, there is a simple remedy: let $r_L^2 = 0$ if solution of (5.66) does not exist. This problem was seen in previous tasks and stems from the fact that the power function does not approach 1 when the alternative goes away from the null value. Indeed, the left limit of (5.66) as a function of ρ^2 is

$$F(r^2; \rho^2 = 0) = \frac{\Gamma\left(\frac{n-1}{2}\right)}{\Gamma\left(\frac{n-p-1}{2}\right)\Gamma\left(\frac{p}{2}\right)} \int_0^{r^2} s^{\frac{p-2}{2}}(1-s)^{\frac{n-p-3}{2}} ds = \mathcal{B}\left(r^2; \frac{p}{2}, \frac{n-p-1}{2}\right),$$

the cdf of the beta distribution. This means the lower-sided CI with confidence level λ does not exist if

$$\mathcal{B}(r^2; p/2, (n-p-1)/2) < \lambda. \tag{5.67}$$

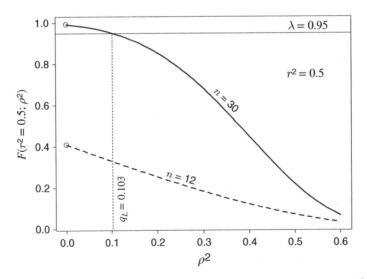

Figure 5.33: *The upper-sided 95% CI does not exist for the obseved $r^2 = 0.5$ when $n = 12$ and $p = 6$. However, it does exist when $n = 30$ and $p = 6$.*

As suggested in Section 4.4, we let $r_L^2 = 0$: if (5.67) takes place the upper-sided CI fails. If the reverse is true we suggest using Newton's iterations for r_L^2 with the adjustment $(F(r^2; r_L^2) - \lambda)/F'(r^2; r_L^2)$, where F' is computed as $F(x; \rho^2) = \int_0^x f'(s, \rho^2)ds$.

The lower-sided CI $(0, r_U^2)$ often exists as the solution to the equation $F(r^2; r_U^2) = 1 - \lambda$, when λ is relatively high, such as $\lambda = 0.95$, because $\lim_{\rho^2 \to 1} F(r^2; \rho^2) = 0$ for $r^2 > 0$. The equations for the upper and lower confidence limits are solved by Newton's algorithm with derivatives of F provided in Proposition 5.15.

Computation of five CIs for ρ^2 is implemented in the R function `twoCIr2` and simulations are in the function `twoCIr2_sim`. The former function has an option of plotting F against ρ^2 given r^2 controlled by parameter `see`. The five CIs are lower-sided, upper-sided, equal-tail, unbiased, and MBESS. All of them have a coverage probability close to nominal. Following is the output for $r^2 = 0.8$, $n = 10$, and $p = 3$.

```
twoCIr2(r2=0.8,n=10,p=3)
                 Lower      Upper
Lower-sided 0.00000000 0.9043361
Upper-sided 0.19227896 1.0000000
Equal-tail  0.06760617 0.9245556
Unbiased    0.08376143 0.9271354
MBESS       0.06762782 0.9239188
```

The equal-tail and MBESS intervals are very close.

5.11.4 The two-sided CI on the log scale

The unbiased estimator of MCC can be negative because the deviation of the estimator from the true positive parameter is measured on the additive scale. Instead, we take the advantage of the fact that ρ^2 is positive and measure the length of the CI on the logarithmic scale as $\ln(q_U/q_L)$, as discussed in Section 2.5.1. The optimal CI is found by solving the equations

$$F(r^2; q_L) - F(r^2; q_U) - \lambda = 0, \quad q_L F'(r^2; q_L) - q_U F'(r^2; q_U) = 0. \qquad (5.68)$$

As shown previously, this system can be solved by Newton's algorithm or by applying general algorithms for solving nonlinear equations, such as implemented in the package `nleqslv`, or using penalized optimization algorithms such as using the function `optimize` as a part of base R. The advantage of the latter approach is that it is general and can be applied to the additive or log scale. For example, to compute the short CI on the log scale we minimize

$$\ln(q_U/q_L) + P \times [F(r^2; q_L) - F(r^2; q_U) - \lambda]$$

over q_L and q_U, where P is the penalty coefficient. For large P, the minimizer is close to the solution of the system (5.68). Figure 5.34 illustrates two-sided CIs

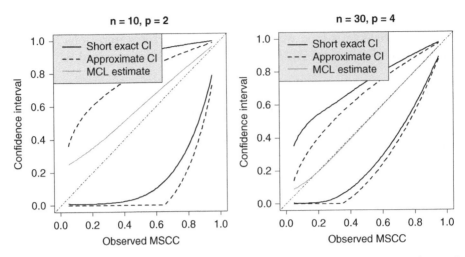

Figure 5.34: *The 95% two-sided CIs for MSCC, the CI on the log scale, and the equal-tail CI using the F-distribution approximation (package MBESS).*

for ρ^2 and was created by calling `moder2_6(N=100,lambda=0.95,PEN=1000)` where `N` is the grid size and `PEN` is the penalty coefficient (P). The lower limit of the equal-tail CI based on the Lee approximation implemented in the package `MBESS` is zero for small r^2. The short CI on the log scale avoids hitting the floor and tends to be above the `MBESS` interval. When the df increases, the difference between intervals becomes less visible.

5.11.5 The MCL estimator

The MCL estimator for ρ^2 is found as the limit point of the system (5.68) when $\lambda \to 0$. Equivalently, the estimator can be defined as

$$r^2_{MCL} = \arg \min_{\rho^2 \in (0,1)} \left(\rho^2 F'(r^2; \rho^2) \right), \qquad (5.69)$$

where r^2 is the observed MCC computed as (5.61), using readily available minimization algorithms, such as `optimize` in R. Alternatively r^2_{MCL} can be found from the equation

$$F'(r^2; \rho^2) + \rho^2 F''(r^2; \rho^2) = 0, \qquad (5.70)$$

where r^2 is the observed MCC. To obtain r^2_{MCL}, we solve equation (5.70) by Newton's algorithm

$$\rho^2_{s+1} = \rho^2_s - \frac{F'(r^2; \rho^2_s) + \rho^2_s F''(r^2; \rho^2_s)}{2F''(r^2; \rho^2_s) + \rho^2_s F'''(r^2; \rho^2_s)}, \quad s = 0, 1, 2, \dots \qquad (5.71)$$

starting from $\rho^2_0 = r^2$. The derivatives of the cdf are found from numerical integration on the interval $(0, r^2)$ over the respective derivatives of the densities as presented in Proposition 5.15.

Theorem 5.16 *The MCL estimator of ρ^2 exists for any observed $0 < r^2 < 1$, $n > 3$.*

Proof. We aim to prove that the minimum of the function $g(u) = uF'(x; u)$ on $u \in (0, 1)$ is attained where $0 < x < 1$ is fixed. Clearly, $\lim_{u \to 0} g(u) = 0$. Now we prove that $\lim_{u \to 1} g(u) = 0$. Indeed, as follows from a representation of F' via integral presented in Proposition 5.15 it is sufficient to see that

$$\lim_{u \to 1} u \left(-\frac{n-1}{2} \frac{1}{1-u} + \frac{r^2 (n-1)^2}{2} \frac{1}{p} \right) (1-u)^{\frac{n-1}{2}}$$

$$= -\frac{n-2}{2} \lim_{u \to 1} (1-u)^{\frac{n-3}{2}} = 0.$$

This also proves that $g(u)$ approaches 0 from below. Therefore we conclude that the minimizer of $g(u)$ is between 0 and 1. □

Figure 5.35 compares the MCL estimator with the unbiased Olkin-Pratt estimator for $n = 10, p = 3$, and $n = 50, p = 7$. This graph was created by calling the function `moder2_3.1(n=10,p=3,N=100)`, where N is the grid size (the package hypergo must be installed beforehand). The R function `optimize` was used to

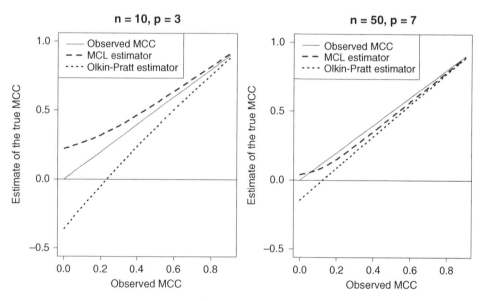

Figure 5.35: *Three estimators of ρ^2: the traditional (observed) MSCC r^2, the MCL estimator, and the unbiased Olkin-Pratt estimator as functions of r^2. For $n = 10$ and $p = 3$, the Olkin-Pratt estimator is negative, when $r^2 < 0.23$. When $n = 20$ and $p = 7$, this estimator is negative when $r^2 < 0.33$. This graph was created by calling the function* moder2_3.1.

find the MCL estimate of r^2 as the minimizer of (5.69). When df increases the three estimators converge. Note that the latter estimator is negative if $r^2 < 0.23$ in the first and if $r^2 < 0.18$ in the second setup. The MCL estimator is always positive.

5.12   Coefficient of determination for linear model

The coefficient of determination (CoD) is a popular quality-of-fit measure of regression analysis (Neter et al. 1989; Draper and Smith 1998). Routinely, it is interpreted as the proportion of variance of the dependent variable explained by predictors, Searle (1971). The CoD is reported as the output summary statistic of linear models in many statistical software packages including R (`lm`), SAS (`PROC REG`), and STATA (`regress`). Despite its importance and widespread popularity the exact statistical inference, including hypothesis testing and CI, is nonexistent. Moreover, an unbiased estimate, the "adjusted R-squared," as a part of the `lm` function in R can take negative values, especially for a small df. The goal of this section is to develop the MCL estimator and an unbiased statistical test.

The CoD is computed in the framework of linear model with fixed/nonrandom predictors,

$$\mathbf{y} = \alpha\mathbf{1} + \mathbf{X}\boldsymbol{\beta} + \varepsilon,$$

where $\mathbf{y}$ is the $n \times 1$ vector of the dependent variable, α is the intercept, $\mathbf{X}$ is the is the $n \times p$ matrix of fixed predictors (independent/explanatory variables), $\boldsymbol{\beta}$ is the $p \times 1$ vector of the slope coefficients, and ε is the random error term distributed as multivariate normal, $\varepsilon \sim \mathcal{N}(0, \sigma^2\mathbf{I})$ under the assumption that matrix $[\mathbf{1}, \mathbf{X}]$ has full rank $(n > p + 1)$. The CoD is defined as

$$R^2 = 1 - \frac{\|\mathbf{y} - \widehat{\mathbf{y}}\|^2}{\|\mathbf{y} - \overline{y}\mathbf{1}\|^2},$$

where $\widehat{\mathbf{y}} = \widehat{\alpha}\mathbf{1} + \mathbf{X}\widehat{\boldsymbol{\beta}}$ is the vector of the predicted values and $\widehat{\alpha}$ and $\widehat{\boldsymbol{\beta}}$ are the least squares (LS) estimates of the intercept and slope coefficients, respectively. Without loss of generality, we will assume that matrix $\mathbf{X}$ is centered meaning $\mathbf{X1} = \mathbf{0}$. Under this assumption, the LS slope coefficients $\widehat{\boldsymbol{\beta}}$ can be derived as $\widehat{\boldsymbol{\beta}} = (\mathbf{X}'\mathbf{X})^{-1}\mathbf{X}'\mathbf{y}$. The adjusted/unbiased CoD has gained much popularity because it reduces the bias and does not necessarily increase with the addition of new predictors (Draper and Smith 1998, p. 140):

$$R^2_{\text{adj}} = 1 - (1 - R^2)\frac{n - 1}{n - p - 1}.$$

Unfortunately, this measure may be negative when df, $n - p - 1$, are small. In this respect, the situation is similar to the previously discussed square multiple correlation coefficient where the unbiased estimator becomes negative for small df. Shortly, we suggest the solution to this problem by abandoning unbiasedness and turning to maximum concentration statistics on the log scale.

5.12.1   CoD and multiple correlation coefficient

There is widespread confusion between CoD and MCC (5.61) including Wikipedia. Many authors mistakenly equate R^2 with the multiple correlation coefficient without recognizing that r^2 applies to random normally distributed predictors and R^2 applies to a linear model, where matrix $\mathbf{X}$ is fixed (nonrandom).

The cdf of R^2 can be expressed via the noncentral F-distribution because the test statistic of the general linear hypothesis under the null is distributed as noncentral F (Demidenko 2020, p. 658)

$$F(x; R_{\mathrm{P}}^2) = \Pr(R^2 \le x) = \mathcal{F}\left(\frac{x}{1-x}\frac{n-p-1}{p}; p, n-p-1, \delta = \frac{nR_{\mathrm{P}}^2}{1-R_{\mathrm{P}}^2}\right),$$
(5.72)

where δ is the noncentrality parameter, and

$$R_{\mathrm{P}}^2 = \frac{1}{1 + n\sigma^2\|\mathbf{X}\boldsymbol{\beta}\|^{-2}}$$
(5.73)

is referred to as the parent/true coefficient of determination (Cramer 1987). It is well known that R^2 has the central F-distribution under the null hypothesis $R_{\mathrm{P}}^2 = 0$, or equivalently when all slope coefficients are equal to zero.

The notion of the true CoD (fixed/nonrandom predictors) is not obvious, because the true/population quantity, sometimes referred to as the parent CoD, is not well defined but can be justified by the following. First, we note that when $n \to \infty$, such that $\|\mathbf{X}\boldsymbol{\beta}\|^2/n \to A > 0$, the cdf of R^2 is a stepwise function: $\Pr(R^2 \le x) = 0$ for $x < R_{\mathrm{P}}^2$ and $\Pr(R^2 \le R_{\mathrm{P}}^2) = 1$, where $R_{\mathrm{P}}^2 = 1/(1 + \sigma^2/A)$ is the limit parent CoD. This justifies the definition (5.73) as the parent CoD. Second, when the matrix $\mathbf{X}$ is random and composed of n samples from the normal distribution, as in the previous section, we have

$$p \lim_{n\to\infty} n\|\mathbf{X}\boldsymbol{\beta}\|^{-2} = p \lim_{n\to\infty} \frac{1}{\boldsymbol{\beta}'(\mathbf{X}'\mathbf{X}/n)\boldsymbol{\beta}} = \frac{1}{\boldsymbol{\beta}'\boldsymbol{\Omega}_x\boldsymbol{\beta}},$$

where $\boldsymbol{\Omega}_x$ is the $p \times p$ population covariance matrix. Then using well-known formulas for the conditional mean (regression) of the multivariate distribution we obtain that $\mathrm{plim}_{n\to\infty}R^2 = \rho^2$, which connects the definition of R_{P}^2 with ρ^2. In short, the statistical inference for R_{P}^2 and ρ^2 is close for large sample sizes but different otherwise.

The difference between distributions of R^2 and r^2 is illustrated in Figure 5.36. When $n \to \infty$, the two distributions converge, but they may be quite different for small df.

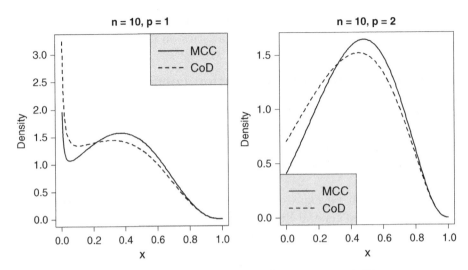

Figure 5.36: *Density functions of the MCC (r^2) and CoD (R^2) for several n and p. For large degree of freedom, the distributions of r^2 and R^2 are close but for small n they are quite different.*

5.12.2   Unbiased test

We aim to develop an exact unbiased test for the null hypothesis $H_0 : R_{\mathrm{P}}^2 = R_{\mathrm{P0}}^2$ along the lines of our derivation presented earlier. Despite the widespread use of the CoD, no two-sided test (not even an equal-tail test) exists today, to the best of our knowledge. The unbiased test rejects this hypothesis if the observed CoD falls outside the interval (R_L^2, R_U^2), where R_L^2 and R_U^2 are found from a system of equations

$$F(R_L^2; R_{\mathrm{P0}}^2) - F(R_U^2; R_{\mathrm{P0}}^2) - \lambda = 0, \; F'(R_L^2; R_{\mathrm{P0}}^2) - F'(R_U^2; R_{\mathrm{P0}}^2) = 0, \quad (5.74)$$

where F' denotes the cdf derivative with respect to R_{P}^2. The size of this test is $\alpha = 1 - \lambda$ and the derivative of the power function at R_{P0}^2 is zero. To facilitate computation of the derivative with respect to R_{P0}^2 an alternative representation via a series of beta cdfs is used (Cramer 1987):

$$F(x; R_{\mathrm{P}}^2) = e^{-\delta/2} \sum_{j=0}^{\infty} \frac{\delta^j}{j! 2^j} B_j, \quad (5.75)$$

where the noncentrality parameter δ is defined as in (5.72) and

$$B_j = \mathcal{B}\left(x; \frac{p}{2} + j, \frac{n - p - 1}{2}\right)$$

is the cdf of the beta distribution. To solve system (5.74), Newton's algorithm or a numerical software package, such as the R package `nleqslv` can be used

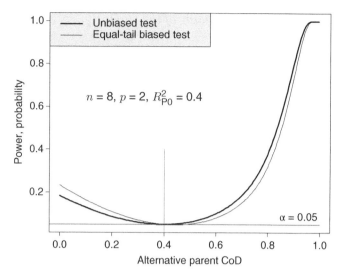

Figure 5.37: *The power functions for the unbiased and equal-tailed tests for the coefficient of determination. The equal-tailed test is slightly biased. The power of detecting the alternative to the left of the null value is small. This graph was created with the function* linr2_test.

starting from the equal-tail solution defined from the $\alpha/2$th and $(1-\alpha/2)$th quantiles of the noncentral F-distribution with df p and $n-p-1$, and noncentrality parameter $\delta = nR_{\mathrm{P}0}^2/(1-R_{\mathrm{P}0}^2)$. The argument Nterm in the R functions that follow defines the number of elements in the sum (5.73). When R_{P}^2 is close to 1, it requires a large number of terms.

The power function of this test is given by

$$P(R_{\mathrm{P}}^2) = 1 + F(R_L^2; R_{\mathrm{P}}^2) - F(R_U^2; R_{\mathrm{P}}^2).$$

As an example, two power functions for testing the parent CoD $R_{P0}^2 = 0.4$ with $n = 8$ and $p = 2$ are depicted in Figure 5.37. The equal-tail test is biased (the derivative of the power function at $R_{P0}^2 = 0.4$ is negative).

5.12.3 The MCL estimator for CoD

The goal of this section is to derive the maximum concentration estimator for the CoD on the log scale, the MCL estimator. Remember that the log scale prevents the estimate from being negative. As follows from Section 2.5.1, the MCL estimator minimizes $R_{\mathrm{P}}^2 F'(R^2; R_{\mathrm{P}}^2)$: that is,

$$R_{MCL}^2 = \arg\min_{0 \le \delta \le 1} \delta F'(R^2; \delta),$$

where the cdf F is given by (5.72) and R^2 is the observed CoD.

Proposition 5.17 *The MCL estimator of the CoD, R^2_{MCL}, exists and is positive.*

Proof. Define a function $g(\delta) = \delta F(R^2; \delta)$, where $0 < x < 1$ and R^2 is fixed from the interval $(0, 1)$. Obviously, $g(0) = 0$. From (5.75), find the derivative

$$F'(x; \delta) = e^{-\delta/2} \left(-\frac{1}{2}(B_0 - B_1) + \sum_{j=2}^{\infty} \frac{\delta^{j-1}}{(j-1)! 2^j} \left(1 - \frac{1}{2}\delta \right) B_j \right).$$

The derivative at zero is negative:

$$\left. \frac{dg}{d\delta} \right|_{\delta=0} = \left. \frac{dF'}{d\delta} \right|_{\delta=0} = -\frac{1}{2}(B_0 - B_1) < 0$$

because the beta cdf is a decreasing function of the shape parameter and therefore $B_0 > B_1$. Next we observe that

$$\lim_{\delta \to \infty} g(\delta) = \lim_{\delta \to \infty} \delta F'(R^2; \delta) = 0,$$

which follows from $\lim_{\delta \to \infty} \delta e^{-\delta/2} = 0$. This means the minimum of $g(\delta)$ exists and is obtained for $\delta > 0$. $\square$

Figure 5.38 depicts the MCL and the adjusted CoD as functions of the observed CoD (R^2) for $p = 3$ and $n = 10$, 20 (minimization is carried out by the built-in R function `optimize`). As follows from this figure, for $n = 10$, the adjusted CoD is negative when R^2 is less than around 0.2, and for $n = 20$ when R^2 is less than around 0.1, but the MCL CoD is always positive. This graph was

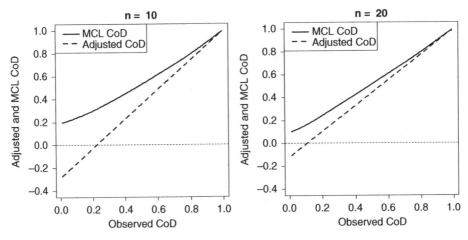

Figure 5.38: *MCL and adjusted CoD, R^2_{adj}, as functions of the observed CoD for $p = 3$ with $n = 10$ and $n = 20$. Thanks to the log scale, the MCL estimate is always positive. This graph was created by function* `linr2_5`.

created by issuing the function `linr2_5(p=3,n=6,N=200,Nterm=500)`, where the argument `Nter` specifies the number of terms in the infinite sum (5.75), and `N` specifies the grid size. It may require several runs to find the optimal `Nterm`. The most considerable difference between the two estimates is when the observed CoD is small.

References

[1] Agresti, A. and Coul, B.A. (1998). Approximate is better than "exact" for interval estimation of binomial probability. *The American Statistician* 52: 119–126.

[2] Algina, J. and Olejnik, S. (2000). Determining sample size for accurate estimation of the squared multiple correlation coefficient. *Multivariate Behavioral Research* 35: 119–136.

[3] Anderson, T.W. (2003). *An Introduction to Multivariate Statistical Analysis*, 3e. New York: Wiley.

[4] Arnold, B.C. (2015). *Pareto Distributions*, 2e. Boca Raton: Chapman & Hall.

[5] Browne, R.H. (2010). The t-test p value and its relationship to the effect size and P(X>Y). *The American Statistician* 64: 30–33.

[6] Clopper, C.J. and Pearson, E.S. (1934). The use of confidence and fiducial limits illustrated in the case of the binomial. *Biometrika* 26: 404–413.

[7] Cohen, A.C. (1951). Estimating parameters of logarithmic-normal distributions by maximum likelihood. *Journal of the American Statistical Association* 46: 206–212.

[8] David, F.N. (1938). *Tables for the Correlation Coefficient*. London: Cambridge University Press.

[9] Demidenko, E. (2013). *Mixed Models. Theory and Applications with R*. Hoboken, NJ: Wiley.

[10] Demidenko, E. (2016). The p-value you can't buy. *The American Statistician* 70: 33–38.

[11] Demidenko, E. (2020). *Advanced Statistics with Applications in R*. Hoboken, NJ: Wiley.

[12] DerSimonian, R. and Laird, N.M. (1986). Meta-analysis in clinical trials. *Controlled Clinical Trials* 7: 177–188.

[13] Draper, N.R. and Smith, H. (1998). *Applied Regression Analysis*, 3e. New York: Wiley.

[14] Fieller, E. (1932). A numerical test of the adequacy of A.T. McKay's approximation. *Journal of the Royal Statistical Society* 95: 699–702.

[15] Fisher, R.A. (1915). Frequency distribution of the values of the correlation coefficient in samples from indefinitely large population. *Biometrika* 10: 507–521.

[16] Guo, B. and Wang, X. (2018). Control charts for the coefficient of variation. *Statistical Papers* 59: 933–955.

[17] Hasan, M.S. and Krishnamoorthy, K. (2017). Improved confidence intervals for the ratio of coefficients of variation of two lognormal distributions. *Journal of Statistical Theory and Applications* 16: 345–353.

[18] Hotelling, H. (1953). New light on the correlation coefficient and its transforms (with discussion). *Journal of the Royal Statistical Society, ser. B* 15: 193–232.

[19] Hu, X., Jung, A., and Qin, G. (2020) Interval estimation for the correlation coefficient. *The American Statistician* 74: 29–36.

[20] Johnson, N.L., Kotz, S., and Balakrishnan, N. (1994). *Continuous Univariate Distributions*, vol. 1, 2e. New York: Wiley.

[21] Johnson, N.L. and Welch, B.L. (1940). Applications of the noncentral t-distribution. *Biometrika* 31: 262–389.

[22] Keating, J.P. and Tripathi, R.C. (1984). Comparison of some two sample tests for equality of variances. *Communications in Statistics – Theory and Methods* 13: 2691–2706.

[23] Kelley, K. (2007). Constructing confidence intervals for standardized effect sizes: Theory, application, and implementation. *Journal of Statistical Software* 20: 1–24.

[24] Kraemer, H.C. (1973). Improved approximation to the non-null distribution of the correlation coefficient. *Journal of the American Statistical Association* 68: 1004–1008.

[25] Krishnamoorthy, K. (2006). *Handbook of Statistical Distributions with Applications*. Boca Raton: Chapman and Hall/CRC.

[26] Lee, Y.S. (1971). Some results on the sampling distribution of the multiple correlation coefficient. *Journal of the Royal Statistical Society, ser. B* 33: 117–130.

[27] Lee, Y.S. (1972). Tables of upper percentage point of the multiple correlation coefficient. *Biometrika* 59: 175–189.

[28] Lehmann, E.L. and Casella, G. (1998). *Theory of Point Estimation*, 2e. New York: Springer-Verlag.

[29] Lehmann, E.L. and Romano, J.P. (2005). *Testing Statistical Hypothesis*, 3e. New York: Springer.

[30] Lim, T.-S. and Loh, W.-Y. (1996). A comparison of tests of equality of variances. *Computational Statistics & Data Analysis* 22: 287–301.

[31] Malik, H.J. (1966). Exact moments of order statistics from the Pareto distribution. *Scandinavian Actuarial Journal* 3–4: 144–157.

[32] McGraw, K.O. and Wong, S.P. (1992). A common language effect size statistic. *Psychological Bulletin* 111: 361–365.

[33] McKay, A.T. (1932). Distribution of the coefficient of variation and the extended 't' distribution. *Journal of the Royal Statistical Society* 95: 695–698.

[34] Mendoza, J.L., Stafford, K.L., and Stauffer, J.M. (2000). Large-sample confidence intervals for validity and reliability coefficients. *Psychological Methods* 5: 356–369.

[35] Miller, E. (1991). Asymptotic test statistics for coefficients of variation, *Communications in Statistics – Theory and Methods* 20: 3351–3363.

[36] Neter, J., Wasserman, W., and Kutner, M.H. (1989). *Applied Linear Regression Models*, 5e. Cambridge: Cambridge University Press.

[37] Newcombe, R.G. (1998). Two-sided confidence intervals for the single proportion. *Statistics in Medicine* 17: 857–872.

[38] Nie, L., Chen, Y., and Chu, H. (2011). Asymptotic variance of maximum likelihood estimator for the correlation coefficient from a BVN distribution with one variable subject to censoring, *Journal of Statistical Planning and Inference* 141: 392–401.

[39] Niwitpong, S. (2013). Confidence intervals for coefficient of variation of lognormal distribution with restricted parameter space. *Applied Mathematical Sciences* 7: 3805–3810.

[40] Olkin, I. and Pratt, J.W. (1958). Unbiased estimation of certain correlation coefficients. *Annals of Mathematical Statistics* 29: 201–211.

[41] Ortega, J.M. and Rheinboldt, W.C. (2000). *Iterative Solution of Nonlinear Equations in Several Variables*. Philadelphia: SIAM.

[42] Pearson, E.S. (1932). Comparison of A.T. McKay's approximation with experimental sampling results. *Journal of the Royal Statistical Society* 95: 703–704.

[43] Ramachandran, K.V. (1958). A test of variances. *Journal of the American Statistical Association* 53: 741–747.

[44] Rao, C.R. (1973). *Linear Statistical Inference and Its Applications*, 2e. New York: Wiley.

[45] Searle, S.R. (1971). *Linear Models*. New York: Wiley.

[46] Searle, S.R., Casella, G., and McCulloch, C.E. (2006). *Variance Components*, 2e. New York: Wiley.

[47] Shao, J. (2003). *Mathematical Statistics*. New York: Springer.

[48] Steiger, J.H. and Fouladi, R.T. (1992). R2: A computer program for interval estimation, power calculation, and hypothesis testing for the squared multiple correlation. *Behavior Research Methods, Instruments and Computers* 4: 581–582.

[49] Vangel, M.G. (1996). Confidence intervals for a normal coefficient of variation. *The American Statistician* 50: 21–26.

[50] Wilks, S.S. (1932). Certain generalizations in the analysis of variance. *Biometrika* 24: 471–494.

[51] Wilson, E.B. (1927). Probable inference, the law of succession, and statistical inference. *Journal of the American Statistical Association* 22: 209–211.

[52] Wolfe, D.A. and Hogg, R.V. (1971). On constructing statistics and reporting data. *American Statistician* 25: 27–30.

[53] Zhou, Y.J., Ho, Z.Y., Liu, J.H., and Tong, T.J. (2020). Hypothesis testing for normal distributions: A unified framework and new developments. *Statistics and Interface* 13: 167–179.

[54] Zou, G.Y. (2007). Exact confidence interval for Cohen's effect size is readily available. *Statistics in Medicine* 26: 3054–3056.

Chapter 6

Multidimensional parameter

This chapter aims to generalize M-statistics to multidimensional parameter $\boldsymbol{\theta} \in R^m$. First, we start with exact tests given the observed statistic value provided the density function is known up to a finite number of unknown parameters. Two versions are developed: the density level (DL) test and the unbiased test, as generalizations of the respective tests for the one-dimensional parameter from Section 2.3 and Section 3.1, respectively. The results related to the Neyman-Pearson lemma presented in Section 1.4 are used to prove the optimality of the DL test. Second, exact confidence regions, dual to the respective tests, are offered. Similarly to the single-parameter case, the point estimator is derived as a limit point of the confidence region when the confidence level approaches zero. A connection between maximum likelihood (ML) and mode estimators (MO) is highlighted. Four examples illustrate our exact statistical inference: simultaneous testing and confidence region for the mean and standard deviation of the normal distribution, two shape parameters of the beta distribution, two-sample binomial problem, and nonlinear regression. The last problem illustrates how an approximate profile statistical inference can be applied to a single parameter of interest.

6.1 Density level test

We aim to generalize an exact test from Section 2.3 with type I error α for the multivariate parameter $\boldsymbol{\theta} \in R^m$ with the simple null hypothesis $H_0 : \boldsymbol{\theta} = \boldsymbol{\theta}_0$ versus the alternative $H_A : \boldsymbol{\theta} \neq \boldsymbol{\theta}_0$ given the continuous, positive, and unimodal pdf $f(\cdot; \boldsymbol{\theta}_0)$ of the p-dimensional statistic $\mathbf{S}$, where $p \geq m$. Remember that for the one-dimensional parameter ($p = m = 1$), we operated with cdf, now we deal with pdf. The dimensions of $\mathbf{S}$ and $\boldsymbol{\theta}$ may differ if not stated otherwise.

M-statistics: Optimal Statistical Inference for a Small Sample, First Edition. Eugene Demidenko.
© 2023 John Wiley & Sons, Inc. Published 2023 by John Wiley & Sons, Inc.

Given the observed value of statistic $\mathbf{S}$, the DL test with the exact type I error $\alpha = 1 - \lambda$ rejects the null if $f(\mathbf{S}; \boldsymbol{\theta}_0) < \kappa_0$, where κ_0 is found from the equation,

$$\int_{\{\mathbf{s}: f(\mathbf{s}; \boldsymbol{\theta}_0) \geq \kappa_0\}} f(\mathbf{s}; \boldsymbol{\theta}_0) ds = \lambda. \qquad (6.1)$$

This test is equivalent to the DL test from Section 2.3. Indeed, for $p = m = 1$, the integration domain $\{\mathbf{s} : f(\mathbf{s}; \boldsymbol{\theta}_0) \geq \kappa_0\}$ turns into an interval (q_L, q_U) such that $f(q_L; \theta_0) = f(q_U; \theta_0)$. The equation (6.1) turns into condition $F(q_U; \theta_0) - F(q_L; \theta_0) = \lambda$, which is equivalent to the system (2.8).

The DL test is optimal in the following sense: as follows from Lemma 1.7, the volume of the acceptance region $\{\mathbf{s} : f(\mathbf{s}; \boldsymbol{\theta}_0) \geq \kappa_0\}$ is minimal among all tests as functions of $\mathbf{S}$ having the same size α.

The p-value of the test equals

$$p = 1 - \int_{\{\mathbf{s}: f(\mathbf{s}; \boldsymbol{\theta}_0) \geq f(\mathbf{S}; \boldsymbol{\theta}_0)\}} f(\mathbf{s}; \boldsymbol{\theta}_0) d\mathbf{s}.$$

This formula is equivalent to the alternative definition of the p-value for the one-dimensional parameter as indicated in Corollary 4.5 of Section 4.1.

Now we turn our attention to solving (6.1) for κ_0. A satisfactory approximation of the left-hand side of (6.1) is $1 - \kappa_0/M$, where $M = \max_{\mathbf{s}} f(\mathbf{s}; \boldsymbol{\theta}_0)$, which turns into exact when the density $f(\mathbf{s}; \boldsymbol{\theta}_0)$ above κ_0 is tent-shaped. The quasi-Newton's algorithm (Ortega and Rheinboldt 2000) to solve equation (6.1) for κ_0 applies:

$$\kappa_{0,k+1} = \kappa_{0k} + M \left(\int_{\{\mathbf{s}: f(\mathbf{s}; \boldsymbol{\theta}_0) \geq \kappa_{0k}\}} f(\mathbf{s}; \boldsymbol{\theta}_0) ds - \lambda \right), \quad k = 0, 1, 2, \ldots \qquad (6.2)$$

starting from $\kappa_{00} = M(1 - \lambda)$, although for some distributions, $\kappa_{00} = 0$ is a better start. Since f is continuous and positive the left-hand side of (6.1) is a decreasing function of κ_0, and therefore there exists a unique solution. The original Newton's algorithm requires the computation of

$$V(\kappa) = \frac{d}{d\kappa} \int_{\{\mathbf{s}: f(\mathbf{s}; \boldsymbol{\theta}_0) \geq \kappa\}} f(\mathbf{s}; \boldsymbol{\theta}_0) ds = \int_{\{\mathbf{s}: f(\mathbf{s}; \boldsymbol{\theta}_0) \geq \kappa\}} ds,$$

the area/volume defined by $\{\mathbf{s} : f(\mathbf{s}; \boldsymbol{\theta}_0) \geq \kappa\}$. This result is closely related to Stokes' theorem (Pfeffer 1987; Spivak 2018) and can be interpreted as the fundamental theorem of calculus in the multidimensional space. Then iterations take the form

$$\kappa_{0,k+1} = \kappa_{0k} + \frac{1}{V_k} \left(\int_{\{\mathbf{s}: f(\mathbf{s}; \boldsymbol{\theta}_0) \geq \kappa_{0k}\}} f(\mathbf{s}; \boldsymbol{\theta}_0) ds - \lambda \right), \quad k = 0, 1, 2, \ldots, \qquad (6.3)$$

where $V_k = V(\kappa_{0k})$.

The power function of the DL test is given by

$$P(\boldsymbol{\theta}; \boldsymbol{\theta}_0) = 1 - \int_{\{\mathbf{s} : f(\mathbf{s}; \boldsymbol{\theta}) \geq \kappa_0\}} f(\mathbf{s}; \boldsymbol{\theta}) d\mathbf{s}, \qquad (6.4)$$

where κ_0 is found from (6.1). Again, this formula is equivalent to the power function (2.11) for the one-dimensional parameter. The following example illustrates the application of (6.1).

Example 6.1 *The DL test for normal mean with unit variance.*

We have $Y_i \overset{iid}{\sim} \mathcal{N}(\mu, 1)$, $i = 1, 2, ..., n$. The joint density is given by $f(\mathbf{y}; \mu_0) = (2\pi)^{-n/2} e^{-0.5\|\mathbf{y} - \mu_0 \mathbf{1}\|^2}$. Since $\{\mathbf{y} : f(\mathbf{y}; \mu_0) \geq \kappa_0\}$ is equivalent to $\{\mathbf{y} : \|\mathbf{y} - \mu_0 \mathbf{1}\|^2 < \eta_0\}$ equation (6.1) takes the form $\Pr(\|\mathbf{Y} - \mu_0 \mathbf{1}\|^2 < \eta_0) = \lambda$, where $\mathbf{Y} \sim \mathcal{N}(\mu_0 \mathbf{1}, \mathbf{I})$. The random variable $\|\mathbf{Y} - \mu_0 \mathbf{1}\|^2$ has a chi-square distribution with n df. The acceptance region of the DL test with size $\alpha = 1 - \lambda$ takes the final form $\sum (Y_i - \mu_0)^2 < C_n^{-1}(\lambda)$, where $C_n^{-1}(\lambda)$ is the λth quantile of the chi-distribution with n df. Note that the DL test differs from the classic likelihood-ratio test as follows from Example 1.3. □

For the one-dimensional parameter $(m = p = 1)$, the set $\{\mathbf{s} : f(\mathbf{s}; \boldsymbol{\theta}_0) \geq \kappa_0\}$ turns into an interval (q_L, q_U) such that $f(q_L; \theta_0) - f(q_U; \theta_0)$. Therefore (6.1) reduces to finding $q_L < q_U$ such that

$$F(q_U; \theta_0) - F(q_L; \theta_0) = \lambda, \quad f(q_L; \theta_0) - f(q_U; \theta_0) = 0,$$

where $\kappa_0 = f(q_L; \theta_0) = f(q_U; \theta_0)$, provided f is unimodal – the same test as in Section 2.3. Although the DL test has the exact type I error, it is biased unless f is symmetric – this is straightforward to see by differentiating the power function evaluated at $\boldsymbol{\theta} = \boldsymbol{\theta}_0$.

Now we comment on the Fisherian paradigm to statistical testing using the p-value. While the concept of the optimal test in the Neyman & Pearson paradigm is well defined via the maximum power, Fisher did not consider the alternative and solely relied on the computation of the p-value and its threshold, such as 0.05, to reject or accept the hypothesis. Unlike the Neyman & Pearson paradigm, the concept of the optimal test is absent in the Fisherian paradigm, or as Lehmann (1993) put it:

"A question that Fisher did not raise was the origin of his test statistic:

Why these rather than some others?"

The DL is an answer to this question as an optimal test having a minimal volume of the acceptance region.

6.2   Unbiased test

To derive a locally unbiased test, we demand that the derivative of the power function at $\boldsymbol{\theta}_0$ must be zero. To construct an unbiased test, we use the same formula as in the DL test, but with a different integration domain and κ_*. Specifically, find κ_* and $\boldsymbol{\theta}_*$ such that

$$\int_{\{s:f(s;\theta_*)\geq\kappa_*\}} f(\mathbf{s};\boldsymbol{\theta}_0)d\mathbf{s} = \lambda, \quad \frac{\partial}{\partial\boldsymbol{\theta}}\int_{\{s_*:f(s;\theta_*)\geq\kappa_*\}} f(\mathbf{s};\boldsymbol{\theta})d\mathbf{s}\bigg|_{\boldsymbol{\theta}=\boldsymbol{\theta}_0} = \mathbf{0}. \quad (6.5)$$

Note that (6.5) defines a system of $1+m$ equations to be solved for $1+m$ unknowns, κ_* and $\boldsymbol{\theta}_*$. Similarly to the DL test, the null hypothesis is rejected if $f(\mathbf{S};\boldsymbol{\theta}_*) < \kappa_*$. The power function of the unbiased test is

$$P(\boldsymbol{\theta};\boldsymbol{\theta}_0) = 1 - \int_{\{s:f(s;\theta_*)\geq\kappa_*\}} f(\mathbf{s};\boldsymbol{\theta})d\mathbf{s},$$

where $\boldsymbol{\theta}_*$ and κ_* depend on $\boldsymbol{\theta}_0$ as the solution of the system (6.5). The p-value of this test is defined as

$$p = 1 - \int_{\{s:f(s;\theta_*)\geq f(S;\theta_*)\}} f(\mathbf{s};\boldsymbol{\theta}_0)d\mathbf{s}$$

By construction, the size of the test is α, and the test is locally unbiased – this is easy to prove by direct differentiation of the power function under integral and appealing to the second equation of (6.5).

Under mild conditions, the test is globally unbiased if the system (6.5) has a unique solution. Indeed, denote

$$\overline{P}_E = \inf_{\|\theta\|\to\infty} P(\boldsymbol{\theta};\boldsymbol{\theta}_0),$$

the lower bound of the power function at infinity. Then, as follows from the Fundamental Theorem of Nonconvex Minimization (Demidenko 2008), if

$$\overline{P}_E > \alpha \quad (6.6)$$

we have $P(\boldsymbol{\theta};\boldsymbol{\theta}_0) > \alpha$ for all $\boldsymbol{\theta} \neq \boldsymbol{\theta}_0$. Since (6.5) has a unique solution the test is globally unbiased because $P(\boldsymbol{\theta}_0;\boldsymbol{\theta}_0) = \alpha$. The condition (6.6) holds for many distributions, particularly when

$$\lim_{a\to\infty} \max_{\|s\|\geq a} f(\mathbf{s};\boldsymbol{\theta}_0) = 0.$$

Under this condition, the test is consistent meaning that $P(\boldsymbol{\theta};\boldsymbol{\theta}_0) \to 1$ when $\|\boldsymbol{\theta} - \boldsymbol{\theta}_0\| \to \infty$.

It is easy to see that for $m = p = 1$, the unbiased test defined by (6.5) turns into the unbiased test (3.1). Indeed, the integration domain in (6.5) turns into an interval $q_L \leq s \leq q_U$ such that $f(q_L; \theta_*) = f(q_U; \theta_*) = \kappa_*$, which implies the following system of three equations for q_L, q_U, and θ_*:

$$F(q_U; \theta_0) - F(q_L; \theta_0) = \lambda, \ f(q_U; \theta_*) - f(q_L; \theta_*) = 0, \ F'(q_U; \theta_0) - F'(q_L; \theta_0) = 0.$$

The solution of this system can be obtained by solving the first and third equations, as in (3.1), and then finding θ_* from the second equation, which is irrelevant. This means that the system (6.5) implies the unbiased test for the one-dimensional parameter (3.1). The condition (6.6) holds: it is a simple consequence of the consistency of the test mentioned in Section 2.3 due to property 4 of Section 2.1.

For the multidimensional location parameter with symmetric density the DL and unbiased tests coincide. Indeed, suppose that $g = g(\mathbf{u})$ is a symmetric density function and $f(\mathbf{s}; \theta) = g(\mathbf{s} - \theta)$. By symmetry we understand that g is symmetric around zero on every line going through the origin, that is, $g(\lambda \mathbf{p}) = g(-\lambda \mathbf{p})$ for every λ and vector $\mathbf{p}$. We aim to show that for $\kappa_* = \kappa_0$ and $\theta_* = \theta_0$, the second equation in (6.5) holds. Indeed, the set $G = \{\mathbf{s} : g(\mathbf{s} - \theta_0) \geq \kappa_*\} = \{\mathbf{s} : f(\mathbf{s}; \theta_0) \geq \kappa_*\}$ is symmetric, and therefore letting $\mathbf{u} = \mathbf{s} - \theta_0$, the following representation holds:

$$\int_{\{\mathbf{s}: f(\mathbf{s}; \theta_0) \geq \kappa_*\}} \frac{\partial}{\partial \theta} f(\mathbf{s}; \theta_0) ds = \int_{\{\mathbf{u}: g(\mathbf{u}) \geq \kappa_*\}} \mathbf{g}'(\mathbf{u}) d\mathbf{u}$$

$$= -\int_{\mathbf{p} \in \partial G} \left(\int_{\lambda \in (-\infty, \infty)} \mathbf{g}'(\lambda \mathbf{p}) d\lambda \right) d\mathbf{p},$$

where $\mathbf{g}'$ is the vector of partial derivatives of g and ∂G is the boundary of G. But due to symmetry $\partial g(\mathbf{s} - \theta)/\partial \theta = -g'(\mathbf{s} - \theta)$, the inner integral vanishes, and therefore the second equation of (6.5) holds for every κ_*. But the first equation dictates $\kappa_* = \kappa_0$, which proves the equivalence of the tests.

6.3 Confidence region dual to the DL test

According to the general definition, the confidence region dual to the DL test is the set of parameters $\theta = \theta_0$ for which the observed value $\mathbf{S}$ belongs to the acceptance region. As follows from Section 6.1, the confidence region for the multidimensional parameter is the set

$$\left\{ \theta \in R^m : \int_{\{\mathbf{s}: f(\mathbf{s}; \theta) \geq f(\mathbf{S}; \theta)\}} f(\mathbf{s}; \theta) ds \leq \lambda \right\} \tag{6.7}$$

with the boundary defined as

$$\left\{ \theta \in R^m : \int_{\{\mathbf{s}: f(\mathbf{s}; \theta) \geq f(\mathbf{S}; \theta)\}} f(\mathbf{s}; \theta) ds = \lambda \right\}. \tag{6.8}$$

The boundary (surface) has dimension $m - 1$ because the m-dimensional parameter is subject to one restriction. For example, for the two-dimensional parameter ($m = 2$), the boundary of the confidence region is a close curve and the confidence region is the area inside the curve. The following example illustrates this definition.

Example 6.2 *Confidence region/interval dual to the DL test for the univariate normal distribution with unit variance.*

We want to illustrate (6.7) and (6.8) by finding the DL confidence region/interval for the mean μ having n independent normally distributed observations $Y_i \sim \mathcal{N}(\mu, 1)$. By letting $S = \overline{Y}$, the pdf f in (6.7) takes the familiar form $f(s; \mu) = (2\pi/n)^{-n/2} e^{-0.5n(s-\mu)^2}$. The integration domain in (6.7) turns into an interval

$$\left\{ s : (s - \mu)^2 \leq (\overline{Y} - \mu)^2 \right\} = \left\{ s : \mu - \left|\overline{Y} - \mu\right| \leq s \leq \mu + \left|\overline{Y} - \mu\right| \right\}.$$

Express the integral as the difference of Φs evaluated at the ends of the previous interval. Then the inequality in (6.7) turns into $\Phi\left(\sqrt{n}\left|\overline{Y} - \mu\right|\right) - \Phi\left(-\sqrt{n}\left|\overline{Y} - \mu\right|\right) \leq \lambda$. After some obvious manipulations, we arrive at an equivalent inequality $\Phi(\sqrt{n}\left|\overline{Y} - \mu\right|) \leq (1 + \lambda)/2$, so that the boundary (6.8) DL CI for μ has the familiar limit points $\overline{Y} \pm \Phi^{-1}((1 + \lambda)/2)/\sqrt{n}$.

Theorem 6.3 *Confidence region with the boundary defined by (6.8) has the exact coverage probability λ.*

Proof. We aim to prove that the random set defined by (6.7) covers the true $\boldsymbol{\theta}$ with probability λ. It is equivalent to proving that for every $\boldsymbol{\theta}$, we have

$$\Pr\left(\int_{\{s:f(s;\boldsymbol{\theta}) \geq f(\mathbf{S};\boldsymbol{\theta})\}} f(\mathbf{s};\boldsymbol{\theta})ds \leq \lambda\right) = \lambda, \tag{6.9}$$

where the pdf of $\mathbf{S}$ is $f(\mathbf{s}; \boldsymbol{\theta})$. Now we realize that the integral inside the parentheses can be expressed as the conditional probability $\Pr\left(f(\mathbf{S}_*; \boldsymbol{\theta}) \geq f(\mathbf{S}; \boldsymbol{\theta}) | \mathbf{S}\right)$, where $\mathbf{S}_*$ has the same distribution as $\mathbf{S}$ independent of $\mathbf{S}$. Introduce two independent identically distributed random variables $X = f(\mathbf{S}_*; \boldsymbol{\theta})$ and $Y = f(\mathbf{S}; \boldsymbol{\theta})$ with the common cdf F. Then the left-hand side of (6.9) can be expressed as $\Pr(\Pr(X \geq Y | Y) \leq \lambda)$. But $\Pr(X \geq Y | Y) = 1 - F(Y)$ since $F(X)$ and $F(Y)$ have the uniform distribution on $(0, 1)$ so the probability that the confidence region (6.7) covers $\boldsymbol{\theta}$ is

$$\Pr(\Pr(X \geq Y | Y) \leq \lambda) = \Pr(1 - F(Y) \leq \lambda) = \Pr(F(Y) \geq 1 - \lambda)$$

$$= 1 - (1 - \lambda) = \lambda$$

as we intended to prove.   □

When $m = 1$ (6.8) defines two points on the line corresponding to the lower and the upper confidence limits, θ_L and θ_U. When $m = 2$, it defines a contour (a closed curve).

As an example, we aim to demonstrate that (6.8) turns into a CI dual to the DL test when $m = 1$. Since f is unimodal for every θ, the equation $f(s; \theta) = f(S; \theta)$ has two roots with one of the roots being S. Let the two roots be $s_1 < s_2 = S$. Then (6.8) implies

$$F(S; \theta_L) - F(s_1; \theta_L) = \lambda, \; f(s_1; \theta_L) = f(S; \theta_L). \qquad (6.10)$$

The solution of this system yields θ_L and s_1. Now consider $S = s_1 < s_2$. Then

$$F(s_2; \theta_U) - F(S; \theta_U) = \lambda, \; f(s_2; \theta_U) = f(S; \theta_U) \qquad (6.11)$$

yields θ_U and s_2. The interval (θ_L, θ_U) covers the true θ with probability λ. As a word of caution: the system defined by (6.10) and (6.11) is not equivalent to the unbiased CI defined by the system (3.4). Qualitatively, they are different in the way the system $f(s_1; \theta_1) = f(s_2; \theta_2)$ is solved. For unbiased CI, the solution is $f(S; \theta_L) = f(S; \theta_U)$, and for the DL CI it is $f(S; \theta_L) = f(s_1; \theta_L)$ and $f(S; \theta_U) = f(s_2; \theta_U)$. The following example clarifies the point.

Example 6.4 *Dual CI for normal variance.*

The DL test for normal variance was derived in Example 2.14 with the acceptance rule $q_L < S/\sigma_0^2 < q_U$, where the quantiles q_L and q_U are found by solving the system (2.12). The dual CI for σ^2 can easily be found directly by solving the acceptance rule/interval for the true σ_0^2 as $\sigma_L^2 = S/q_U$ and $\sigma_U^2 = S/\sigma_L^2$. Since $\Pr(q_L < S/\sigma_0^2 < q_U) = \lambda$ by inversion for σ_0^2, we infer that $\Pr(\sigma_L^2 < \sigma_0^2 < \sigma_U^2) = \lambda$. Now we demonstrate that this CI can be found from the general system of equations (6.10) and (6.11). Indeed, the first system equivalently is rewritten as

$$C_{n-1}(S/\sigma_L^2) - C_{n-1}(s_1/\sigma_L^2) = \lambda, \; c_{n-1}(S/\sigma_L^2) - c_{n-1}(s_1/\sigma_L^2) = 0.$$

After letting $q_U = S/\sigma_L^2$ and $q_L = s_1/\sigma_L^2$, we solve the system

$$C_{n-1}(q_U) - C_{n-1}(q_L) = \lambda, \; (n-3)\ln(q_U/q_L) - (q_U - q_L) = 0$$

as before (2.12) with the solution $\sigma_L^2 = S/q_U$. The system (6.11) turns into a similar system but σ_L^2 replaced with σ_U^2, which implies $\sigma_U^2 = S/q_L$. Thus, the dual CI for σ^2 takes the form $(S/q_U, S/q_L)$. Note that the second equation of the unbiased CI given by the system (1.1) is different – it uses the coefficient $(n - 1)$ instead of $(n - 3)$. Briefly, equations (6.10) and (6.11) do not produce an unbiased CI for normal variance. Consequently, (6.8) does not produce an unbiased confidence region in the general case. $\qquad \square$

Now, from a straightforward adaptation of Definition 3.8, we derive a new DL estimator as the limit point of the confidence region (6.8) when the confidence level approaches zero. Indeed, when $\lambda \to 0$, the confidence region converges to the point in the parameter space where the density $f(s; \boldsymbol{\theta})$ reaches its maximum over $\mathbf{S}$ given $\boldsymbol{\theta}$. Therefore the following definition.

Definition 6.5 *Density level estimator. When the dimension of statistic $\mathbf{S}$ and the parameter space is the same $(m = p)$, the DL estimator, $\widehat{\boldsymbol{\theta}}_{DL} = \widehat{\boldsymbol{\theta}}_{DL}(\mathbf{S})$, is defined as the solution of the equation*

$$\left. \frac{\partial f(\mathbf{s}; \boldsymbol{\theta})}{\partial \mathbf{s}} \right|_{\mathbf{s} = \mathbf{S}} = \mathbf{0}. \tag{6.12}$$

Note that equal dimension is a necessary condition for the uniqueness and existence of the estimator. In other words, the DL estimator is the value of the parameter for which $\mathbf{S}$ is the mode. We draw the attention of the reader to the difference between the MO and DL estimators defined by the conditions (3.6) and (6.12), respectively. The former yields the maximum of the pdf over the parameter and the latter the maximum over the statistic.

Example 6.6 *DL estimator for normal variance.*

Since the pdf is $f(s; \sigma^2) = \sigma^{-2} c_{n-1}(s/\sigma^2)$ we have

$$\frac{\partial \ln f(s; \sigma^2)}{\partial s} = \left(\frac{n-3}{2} \ln \frac{s}{\sigma^2} - \frac{s}{2\sigma^2} \right)' = \frac{n-3}{2s} - \frac{1}{2\sigma^2} = 0,$$

which yields $\widehat{\sigma}_{DL}^2 = S/(n-3)$. Note that the mode estimator, given by (1.14), uses $n-1$ in the denominator.

6.4 Unbiased confidence region

The goal of this section is to develop an unbiased confidence region dual to the unbiased test from Section 6.2. Following the principle of duality, our strategy is to find the set of $\boldsymbol{\theta} = \boldsymbol{\theta}_0$, called the confidence region, for which the null hypothesis, given the test statistic $\mathbf{S}$, is accepted. The boundary of the unbiased confidence region with exact confidence level λ is defined as a set of $\boldsymbol{\theta} \in R^m$ found from simultaneous solving the following $1 + m$ equations:

$$\int_{\{\mathbf{s}: f(\mathbf{s}; \boldsymbol{\theta}_*) \geq f(\mathbf{S}; \boldsymbol{\theta}_*)\}} f(\mathbf{s}; \boldsymbol{\theta}) d\mathbf{s} = \lambda, \quad \frac{\partial}{\partial \boldsymbol{\theta}} \int_{\{\mathbf{s}: f(\mathbf{s}; \boldsymbol{\theta}_*) \geq f(\mathbf{S}; \boldsymbol{\theta}_*)\}} f(\mathbf{s}; \boldsymbol{\theta}) d\mathbf{s} = 0, \tag{6.13}$$

where the m-dimensional unknown vector $\boldsymbol{\theta}_*$ is treated as a nuisance parameter. This system defines a manifold of dimension $m - 1$ because there are $1 + m$

equations with $2m$ unknowns, $\boldsymbol{\theta}$ and $\boldsymbol{\theta}_*$. For $m = p = 1$, this system defines a pair of confidence limits with two solutions corresponding to θ_L and θ_U. Similarly to (6.7), the unbiased confidence region is the set of $\boldsymbol{\theta} \in R^m$ for which

$$\int_{\{s:f(s;\boldsymbol{\theta}_*)\geq f(\mathbf{S};\boldsymbol{\theta}_*)\}} f(\mathbf{s};\boldsymbol{\theta})d\mathbf{s} \leq \lambda, \qquad (6.14)$$

where $\boldsymbol{\theta}$ and $\boldsymbol{\theta}_*$ satisfy (6.13). The first equation of (6.13) guarantees that the confidence region covers the true parameter with probability λ and the second equation is a necessary condition that the coverage probability attains its maximum at the true parameter (the derivative is zero).

The proof that the unbiased confidence region (6.14) covers the true parameter with exact coverage probability λ follows the line of the proof of Theorem 6.3. Indeed, express (6.14) as the conditional probability $\Pr\left(f(\mathbf{S}_*;\boldsymbol{\theta}_*) \geq f(\mathbf{S};\boldsymbol{\theta}_*)|\mathbf{S};\boldsymbol{\theta}\right)$ and note that the probability is computed at $\boldsymbol{\theta} \neq \boldsymbol{\theta}_*$. However, it does not affect the rest of the proof with $X = f(\mathbf{S}_*;\boldsymbol{\theta}_*)$ and $Y = f(\mathbf{S};\boldsymbol{\theta}_*)$ having the common cdf $F(\cdot;\boldsymbol{\theta})$.

When $\lambda \to 0$, the unbiased confidence region shrinks to the MO estimator. Consequently, when $\mathbf{s}$ is the vector of all observations, it converges to the maximum likelihood estimate. Indeed, since f is unimodal for every $\boldsymbol{\theta}_*$, when $\lambda \to 0$, the set $A = \{\mathbf{s} : f(\mathbf{s};\boldsymbol{\theta}_*) \geq f(\mathbf{S};\boldsymbol{\theta}_*)\}$ converges to point $\mathbf{S}$ with the volume

$$\text{vol}(A) = \int_{\{s:f(s;\boldsymbol{\theta}_*)\geq f(\mathbf{S};\boldsymbol{\theta}_*)\}} d\mathbf{s}.$$

It is a fundamental fact of the multivariable calculus, closely related to Stokes' theorem (Section 6.1), that

$$\lim_{A\to\mathbf{S}} \frac{1}{\text{vol}(A)} \int_{\{s:f(s;\boldsymbol{\theta}_*)\geq f(\mathbf{S};\boldsymbol{\theta}_*)\}} f(\mathbf{s};\boldsymbol{\theta})d\mathbf{s} = f(\mathbf{S};\boldsymbol{\theta}).$$

Thus, the second condition in (6.13) rewrites as

$$\frac{\partial}{\partial\boldsymbol{\theta}} f(\mathbf{S};\boldsymbol{\theta}) = \mathbf{0}$$

with the solution as

$$\widehat{\boldsymbol{\theta}}_{MO} = \arg\max_{\boldsymbol{\theta}} f(\mathbf{S};\boldsymbol{\theta}),$$

the MO estimator of $\boldsymbol{\theta}$. In other words, the MO estimator maximizes the pdf over the parameter evaluated at the observed $\mathbf{S}$. This result is an obvious generalization of the one-dimensional Definition 3.8 of the MO estimator.

When $\mathbf{S}$ comprises all observations, $\mathbf{S} = (Y_1, Y_2, ..., Y_n)$, the M) estimator turns into the maximum likelihood estimator (MLE). This connection between M-statistics and methods of the classical maximum likelihood shed new light on the optimal property of the MLE in small samples discussed earlier in Chapter 3.

Now we illustrate (6.13) with the one-dimensional case, $m = p = 1$ studied in previous chapters. For the one-dimensional case, the integration domain $\{s : f(s; \boldsymbol{\theta}_*) \geq f(S; \boldsymbol{\theta}_*)\}$ turns into an interval (s_1, s_2) under the assumption of unimodality. Note that either s_1 or s_2 equals S for any θ_*. First, consider the case when $s_1 < s_2 = S$. Then system (6.13) turns into three equations:

$$F(S; \theta) - F(s_1; \theta) = \lambda, \ f(s_1; \theta_*) = f(S; \theta_*), \ F'(s_1; \theta) - F'(S; \theta) = 0 \quad (6.15)$$

with three unknowns: s_1, θ, and θ_*, where θ is treated as a confidence limit and s_1 and θ_* are nuisance parameters. Second, for the case when $S = s_1 < s_2$ we arrive at a similar system

$$F(s_2; \theta) - F(S; \theta) = \lambda, \ f(s_2; \theta_*) = f(S; \theta_*), \ F'(s_2; \theta) - F'(S; \theta) = 0 \quad (6.16)$$

with the solution for θ as the second confidence limit and s_2 and θ_* treated as nuisance parameters.

The following example illustrates that for normal variance this system produces an unbiased CI.

Example 6.7 *Normal variance.*

We aim to show that the systems (6.15) and (6.16) imply the unbiased CI for σ^2 derived in our motivating example of Section 1.3. Since $F(s; \sigma^2) = C_{n-1}(s/\sigma^2)$, $F'(s; \sigma^2) = -s/\sigma^4 c_{n-1}(s/\sigma^2)$, and $f(s; \sigma^2) = \sigma^{-2} c_{n-1}(s/\sigma^2)$, the system (6.15), to be solved for s_1, σ_*^2, and σ^2, equivalently is rewritten as

$$C_{n-1}(S/\sigma^2) - C_{n-1}(s_1/\sigma^2) = \lambda, \ c_{n-1}(s_1/\sigma_*^2) = c_{n-1}(S/\sigma_*^2),$$
$$s_1 c_{n-1}(s_1/\sigma^2) - S c_{n-1}(S/\sigma^2) = 0.$$

Change of variables: $q_U = S/\sigma^2$, $q_L = s_1/\sigma^2$. The unknown σ_*^2 can be uniquely expressed through q_L and q_U from the second equation after s_1 is determined, but it is not important at this point. Now divide the third equation by σ^2. Then the system for q_L and q_U takes the form

$$C_{n-1}(q_U) - C_{n-1}(q_L) = \lambda, \ q_L c_{n-1}(q_L) - q_U c_{n-1}(q_U).$$

This is the same system as for the unbiased CI for normal variance given by equations (3.1) and (1.10) that produces the lower confidence limit $\sigma_L^2 = S/q_U$.

When $S = s_1 < s_2$, the system (6.16) takes the form

$$C_{n-1}(s_2/\sigma^2) - C_{n-1}(S/\sigma^2) = \lambda, \ c_{n-1}(s_2/\sigma_*^2) = c_{n-1}(S/\sigma_*^2),$$
$$s_2 c_{n-1}(s_2/\sigma^2) - S c_{n-1}(S/\sigma^2) = 0.$$

Applying a similar change of variables we derive the upper confidence limit $\sigma_U^2 = S/q_L$. Thus we conclude that the interval (σ_L^2, σ_U^2) derived from (6.15) and (6.16) is the same as the unbiased CI derived in Section 1.3.

6.5   Simultaneous inference for normal mean and standard deviation

The organization of this section is as follows: we start by constructing an exact statistical test for the simultaneous null hypothesis $\mu = \mu_0$ and $\sigma = \sigma_0$ versus $\mu \neq \mu_0$ or $\sigma \neq \sigma_0$. Then we discuss an exact confidence region for (σ, μ). Specifically, the goal of this section is to develop an exact unbiased test along with its power function, the exact unbiased confidence region, and the implied DL and MO estimators following the general methodology outlined earlier. Although in the past the simultaneous statistical inference applied to mean and variance, we parametrize the problem via the mean and standard deviation that makes the plot of the confidence region with the axes having the same measurement units.

6.5.1   Statistical test

Simultaneous testing for the mean and variance of the normal distribution has a long history dating back to Neyman and Pearson (1936). Most of authors used the likelihood ratio test in conjunction with asymptotic results leading to the chi-square distribution. Although much literature exists on the two-sample test, surprisingly less effort has gone onto the simpler one-sample test with the simple hypothesis on the mean and variance. Several authors, including Choudhari et al. (2001), derived the exact distribution of the test statistic with the exact type I error.

Given a random sample $\{X_i, i = 1, ..., n\}$ from a normally distributed general population with standard deviation σ and mean μ, we aim to construct an unbiased test for the null hypothesis $H_0 : \mu = \mu_0, \sigma = \sigma_0$ versus an alternative that either $\mu \neq \mu_0$ or $\sigma \neq \sigma_0$. To conduct the test, we use a pair of independent statistics $S_1 = \overline{X}$ and $S_2 = \sum (X_i - \overline{X})^2$ with the joint bivariate pdf

$$f(s_1, s_2; \mu, \sigma) = \frac{\sqrt{n}}{\sigma^3} \phi \left(\frac{\sqrt{n}(s_1 - \mu)}{\sigma} \right) c_{n-1} \left(\frac{s_2}{\sigma^2} \right). \tag{6.17}$$

The tests to be discussed later require evaluation of the double integral over the domain $\{\mathbf{s} : f(\mathbf{s}; \boldsymbol{\theta}) \geq \kappa\}$ where $\mathbf{s} = (s_1, s_2)$ and $\boldsymbol{\theta} = (\mu, \sigma)$. Several strategies can be used to compute the integral

$$\int_{\{(s_1, s_2): f(s_1, s_2; \mu_*, \sigma_*) \geq \kappa_*\}} f(s_1, s_2; \mu, \sigma) ds_1 ds_2. \tag{6.18}$$

The easiest, but slow yet general, approach is to employ the Monte Carlo method by simulating a large number of pairs (s_1, s_2) with the joint pdf f dependent on σ and μ, and then computing the proportion of pairs for which $f(s_1, s_2; \mu_*, \sigma_*) \geq \kappa_*$. The same approach can be used to approximate integrals for derivatives in (6.5).

A faster and more precise but specific approach takes advantage of the fact that the region of integration, given μ_*, σ_*, and κ_*, can be expressed as the set of points (s_1, s_2) defined by the inequalities $a \leq s_1 \leq b$ and $c(s_1) \leq s_2 \leq d(s_1)$, where the integration limits $a < b$ and functions $c \leq d$ are to be determined. Then the double integral is reduced to a single integral by use of the standard normal cdf function Φ as follows:

$$\int_{f(s_1,s_2;\mu_*,\sigma_*)\geq\kappa_*} f(s_1, s_2; \mu, \sigma)ds_1 ds_2 = \int_a^b \left(\int_{c(s_2)}^{d(s_2)} f(s_1, s_2; \mu, \sigma)ds_1 \right) ds_2$$

$$= \sigma^{-2} \int_a^b \left[\Phi(\sqrt{n}(d(s_2) - \mu)/\sigma) - \Phi(\sqrt{n}(c(s_2) - \mu)/\sigma) \right] c_{n-1}(s_2/\sigma^2)ds_2. \quad (6.19)$$

To find the integration limits we solve the equation

$$\ln f(s_1, s_2; \sigma_*, \mu_*) = -n \ln \sigma_* - \frac{n}{2\sigma_*^2}(s_1 - \mu_*)^2 - H + \frac{n-3}{2} \ln s_2 - \frac{1}{2\sigma_*^2}s_2 = \ln \kappa_*,$$

for s_1 and s_2, where

$$H = -0.5 \ln n + 0.5 \ln(2\pi) + \ln 2^{(n-1)/2}\Gamma((n-1)/2).$$

The following result is instrumental for finding the integration limits in (6.19) and can be considered a special case of Newton's algorithm discussed by Ortega and Rheinboldt (2000).

Proposition 6.8 *If* $c > \ln a - 1$ *equation*

$$\ln s - \frac{s}{a} = c, \quad a > 0$$

does not have roots for $s > 0$. *Otherwise, the two roots can be effectively found by the monotonically converging Newton's iterations.*

By differentiation, it is easy to see that the maximum of the left-hand side of the equation to solve attained at $s_{\max} = a$ with the maximum $\ln a - 1$. This means the equation has no roots if $c > \ln a - 1$. When $c = \ln a - 1$, the equation has the unique root $s = a$. Consider the case when $c < \ln a - 1$. Twice differentiation implies that the left-hand side is a concave function of s, and therefore there are two roots on both sides of the maximizer a: that is, $s_L < a < s_R$. To find the starting value for s_L, we use the inequality $\ln s - s/a < \ln s$, which yields $s_{L0} = e^c$. To find the starting value for s_R, we represent

$$\ln s - \frac{s}{a} = \left(\ln s - \frac{s}{2a} \right) - \frac{s}{2a}.$$

Since the minimum value in the square brackets is $\ln(2a) - 1$, we can bound

$$\ln s - \frac{s}{a} < (\ln(2a) - 1) - \frac{s}{2a},$$

which yields the second starting value $s_{R0} = 2a(\ln(2a) - c - 1)$. Since the function $\ln s - s/a$ is convex Newton's algorithm

$$s_{k+1} = s_k - \frac{\ln s_k - s_k/a - c}{1/s_k - 1/a}, \ k = 0, 1, 2, ...$$

monotonically converges to the left and right root when starting from s_{L0} and s_{R0}, respectively. □

The limits of integration, $c(s_1)$ and $d(s_1)$, are defined as

$$c(s_1) = \mu_* - \sigma_*\sqrt{2(A(s_2) - G)/n}, \ d(s_1) = \mu_* + \sigma_*\sqrt{2(A(s_2) - G)/n},$$

where

$$A(s_2) = \frac{n-3}{2}\ln s_2 - \frac{1}{2\sigma_*^2}s_2, \ G = \ln \kappa_* + H + n\ln\sigma_*.$$

As follows from the previous proposition, the limits a and b are found from equation $A(s_2) = G$ and solved by Newton's algorithm:

$$s_{2,k+1} = s_{2,k} + \frac{2(G - A(s_{2k}))}{(n-3)/s_{2k} - 1/\sigma_*^2}, k = 0, 1, 2, ..$$

To compute a, we start from $s_{20} = e^{2G/(n-3)}$. To compute b we start from

$$s_{20} = 2n_3\sigma_*^2 \ln\left(2n_3\sigma_*^2\right) - 2n_3\sigma_*^2 - 4\sigma_*^2 G,$$

where $n_3 = n - 3$. These iterations monotonically converge to a and b, respectively.

To compute the derivatives with respect to σ and μ we use the formula

$$\int_{\{s:f(s;\boldsymbol{\theta})\geq f(S;\boldsymbol{\theta})\}} \frac{\partial}{\partial\boldsymbol{\theta}}f(s;\boldsymbol{\theta})ds = \int_{\{s:f(s;\boldsymbol{\theta})\geq f(S;\boldsymbol{\theta})\}} f(s;\boldsymbol{\theta})\frac{\partial\ln f(s;\boldsymbol{\theta})}{\partial\boldsymbol{\theta}}ds, \qquad (6.20)$$

which simplifies computation, especially when the joint density is the product with separated parameters as in the present example.

Now we provide some details on how to reduce (6.20) to the one-dimensional integral, similarly to (6.19). The following formulas apply:

$$\int_A^B u\phi(u)du = \phi(A) - \phi(B), \ \int_A^B u^2\phi(u)du = (\Phi(B) - \Phi(A)) - (B\phi(B) - A\phi(A)).$$

Obviously,

$$\frac{\partial}{\partial \sigma} \ln f(s_1, s_2; \sigma, \mu) = -\frac{n}{\sigma} + \frac{n(s_1 - \mu)^2 + s_2}{\sigma^3},$$

$$\frac{\partial}{\partial \mu} \ln f(s_1, s_2; \sigma, \mu) = \frac{n(s_1 - \mu)}{\sigma^2}.$$

Specifically, integration over the derivative with respect to σ implies

$$\int_{f(s_1, s_2; \sigma_*, \mu_*) \geq \kappa_*} f(s_1, s_2; \mu, \sigma) \frac{\partial}{\partial \sigma} \ln f(s_1, s_2; \sigma, \mu) ds_1 ds_2$$

$$= \frac{\sqrt{n}}{\sigma^6} \int_a^b \left(\int_{c(s_2)}^{d(s_2)} \left(n(s_1 - \mu)^2 + s_2 - n\sigma^2 \right) \phi \left(\sqrt{n}(s_1 - \mu)/\sigma \right) ds_1 \right)$$

$$\times c_{n-1}(s_2/\sigma^2) ds_2.$$

Letting $D = -n/\sigma + s_2/\sigma^3$ after change of variable $u = \sqrt{n}(s_1 - \mu)/\sigma$, we get

$$\frac{\sqrt{n}}{\sigma} \int_{c(s_2)}^{d(s_2)} \left(D + \frac{n(s_1 - \mu)^2}{\sigma^3} \right) \phi \left(\sqrt{n}(s_1 - \mu)/\sigma \right) ds_1$$

$$= \int_{\sqrt{n}(c(s_2) - \mu)/\sigma}^{\sqrt{n}(d(s_2) - \mu)/\sigma} \left(D + u^2/\sigma \right) \phi(u) du$$

$$= D \int_{\sqrt{n}(c(s_2) - \mu)/\sigma}^{\sqrt{n}(d(s_2) - \mu)/\sigma} \phi(u) du + \frac{1}{\sigma} \int_{\sqrt{n}(c(s_2) - \mu)/\sigma}^{\sqrt{n}(d(s_2) - \mu)/\sigma} u^2 \phi(u) du$$

$$= \left(D + \frac{1}{\sigma} \right) \left[\Phi \left(\frac{\sqrt{n}(d(s_2) - \mu)}{\sigma} \right) - \Phi \left(\frac{\sqrt{n}(c(s_2) - \mu)}{\sigma} \right) \right]$$

$$- \frac{1}{\sigma} \left(\frac{\sqrt{n}(d(s_2) - \mu)}{\sigma} \right) \phi \left(\frac{\sqrt{n}(d(s_2) - \mu)}{\sigma} \right)$$

$$+ \frac{1}{\sigma} \left(\frac{\sqrt{n}(c(s_2) - \mu)}{\sigma} \right) \phi \left(\frac{\sqrt{n}(c(s_2) - \mu)}{\sigma} \right).$$

Now we work on the derivative with respect to μ:

$$\frac{\sqrt{n}}{\sigma} \int_{f(s_1, s_2; \sigma_*, \mu_*) \geq \kappa_*} f(s_1 s_2; \mu, \sigma) \frac{\partial}{\partial \mu} \ln f(s_1, s_2; \sigma, \mu) ds_1 ds_2$$

$$= \frac{\sqrt{n}}{\sigma} \int_a^b \left(\int_{c(s_2)}^{d(s_2)} \left(\frac{n(s_1 - \mu)}{\sigma^2} \right) \phi \left(\sqrt{n}(s_1 - \mu)/\sigma \right) ds_1 \right) c_{n-1}(s_2/\sigma^2) ds_2.$$

Using the same change of variable we obtain

$$
\frac{\sqrt{n}}{\sigma} \int_{c(s_2)}^{d(s_2)} \left(\frac{n(s_1 - \mu)}{\sigma^2} \right) \phi\left(\sqrt{n}(s_1 - \mu)/\sigma \right) ds_1
$$

$$
= \int_{\sqrt{n}(c(s_2)-\mu)/\sigma}^{\sqrt{n}(d(s_2)-\mu)/\sigma} \left(\frac{\sqrt{n}}{\sigma} u \right) \phi\left(u \right) du = \frac{\sqrt{n}}{\sigma} \int_{\sqrt{n}(c(s_2)-\mu)/\sigma}^{\sqrt{n}(d(s_2)-\mu)/\sigma} u \phi\left(u \right) du
$$

$$
= -\frac{\sqrt{n}}{\sigma} \left[\phi\left(\frac{\sqrt{n}(d(s_2) - \mu)}{\sigma} \right) - \phi\left(\frac{\sqrt{n}(c(s_2) - \mu)}{\sigma} \right) \right].
$$

Now we provide some details on the computation of the power function and the respective $\kappa_0, \kappa_*, \sigma_*,$ and μ_*. Refer to Figure 6.1, created by calling `testMS()`. The integral (6.19) is computed by the local function

```
int.an(mu,sigma,mu.star,sigma.star,kappa.star,n)
```

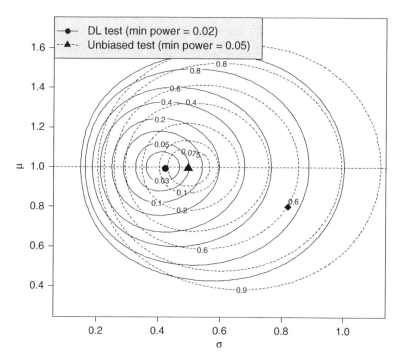

Figure 6.1: *The contour plot of the power functions for the DL test (6.1) and unbiased test (6.5) with $H_0 : \sigma_0 = 0.5, \mu_0 = 1,$ and $n = 10$. The minimum of the power function of the DL test (filled circle) is less than $\alpha = 0.05$. For the unbiased test, the minimum power reaches at the null value (filled triangle). The diamond symbol at around $(0.8, 0.8)$ represents the power value 0.06 obtained from simulations that matches the theoretical power. This graph was created by calling* `testMS()`.

that implements the algorithm described earlier. We aim at testing $\sigma_0 = 0.5$ and $\mu_0 = 1$ with $n = 10$. For the DL test, we solve equation (6.1) by the quasi Newtons's algorithm (6.2), which quickly converges to $\kappa_0 = 0.049$. To verify the computation, we use three other methods: approximation using the maximum of the bivariate density (M), simulations, and the built-in function uniroot. The output is presented here.

```
> testMS()
                        kappa       integral
approximation        0.04758261    0.9513975
simulations          0.05307092    0.9459230
Newton's algorithm   0.04898472    0.9499977
uniroot              0.04897187    0.9500105
quasi-Newton         0.04898197    0.9500004
Time difference of   8.912966 secs
```

Remarkably, the approximation works very well.

For the unbiased test, we solve equations (6.5) by the derivative-free algorithm differential evolution (DE) implemented in the R package DEoptim (Ardia et al. 2011). Alternatively, the R package nleqslv for solving nonlinear equations could be used. Note that the power function of the DL test is smaller than $\alpha = 0.05$ and does not reach its minimum (filled circle) at the null point $(0.5, 1)$. On the the other hand, the minimum of the power function of the unbiased test is always equal or greater than 0.05 and attains its minimum right on the null value, thanks to the second equation of (6.5). To check the power computation, we used simulations at the alternative values $\sigma = 0.82$ and $\mu = 0.8$: see the diamond symbol with the power equal to 0.61 that closely matches the power computed by numerical integration (number of simulations = 100,000).

6.5.2   Confidence region

Several authors have dealt with simultaneous confidence regions for the mean and variance, such as Mood (1950), Wilks (1962), and Arnold and Shavelle (1998), but none offer an exact unbiased test or a test with the minimum-volume acceptance area.

Figure 6.2 displays several confidence regions for $n = 10$ created by the R function confMS(), where σ is displayed on the x-axis and μ on the y-axis. This function has several arguments that allow constructing three confidence regions under different scenarios with the default values as follows: n=10 is the sample size; mu0=1 is the true value for the mean, μ_0; sd0=0.5 is the true value for the SD, σ_0; alpha=0.05 is complementary to the confidence level, $\lambda = 1 - \alpha$; N=100 is the number of points when plotting contours; ss=3 is the seed number, and nSim=100000 is the number of simulations. This function shares several local functions with the previous code testMS, however, the

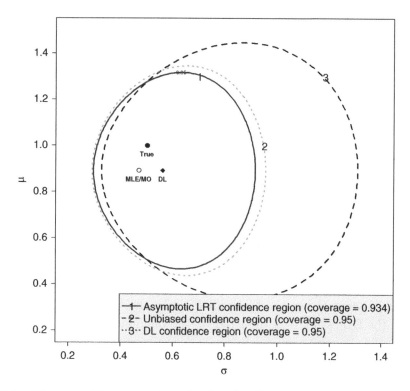

Figure 6.2: *Three confidence regions for (σ, μ). The minimum volume and unbiased confidence regions have the exact coverage probability. The LR and minimum volume confidence regions are close. This graph was created by calling the R function* confMS().

package `nleqslv` is used to solve system (6.13) – it takes about 2.5 minutes to create the graph. Three regions are computed with the randomly generated values $S_1 = \overline{Y} = 0.891217$ and $S_2 = \sum_{i=1}^{n}(Y_i - \overline{Y})^2 = 2.202743$ depending on `ss` from the normal and chi-squared distributions with $\mu_0 = 1$ and $\sigma_0 = 0.5$, respectively. To contrast our confidence region with the traditional approaches the asymptotic likelihood-ratio (LR) confidence region (Arnold and Shavelle 1998) is used as a set of points defined as

$$\left\{ (\sigma, \mu) : n(S_1 - \mu)^2/\sigma^2 + n\ln(n\sigma^2/S_2) + S_2/\sigma^2 - n < q_\lambda \right\},$$

where q_λ is the λth quantile of the chi-square distribution with two df.

To plot the boundary of the unbiased confidence region, we compute κ, using Newton's algorithm (6.2), as a function of μ and σ on the grid of values from the rectangle $[0.2, 1.4] \times [0.2, 1.4]$ and then plot the set of σ and μ such that $f(S_1, S_2; \mu, \sigma) - \kappa(\mu, \sigma) = 0$ using `contour` function in R. Alternatively, we can derive the same confidence boundary/curve by computing the integral by letting $\kappa = f(S_1, S_2; \mu, \sigma)$ and then apply `contour` with the level value equal λ that avoids finding κ. The package `nleqslv` is used to solve the three equations for κ_*, μ_*, and σ_*. See also, Frey et al. (2009) and Zhang (2017) for other optimal exact confidence regions. Interestingly, the DL confidence region is substantially larger but yet covers the true σ_0 and μ_0 with the nominal probability. Another interesting fact is that the LR and unbiased confidence regions are close but the former has a smaller coverage probability.

Finding the MO estimator from (6.17) is elementary:

$$\widehat{\mu}_{MO} = \widehat{\sigma}_{ML} = \overline{Y}, \quad \widehat{\sigma}_{MO} = \widehat{\sigma}_{ML} = \sqrt{\frac{1}{n}\sum_{i=1}^{n}(Y_i - \overline{Y})^2}.$$

Interestingly, the MO estimator for σ in simultaneous inference is different from that derived in Section 5.1.3. Not surprisingly, $\widehat{\sigma}_{MO}$ is the same as the MLE because $\overline{Y}$ and S are sufficient statistics for μ and σ. The DL estimator of σ is the square root of $\widehat{\sigma}_{DL}^2$ found in Example 6.6. The true value of the parameter $\boldsymbol{\theta}_0 = (\sigma_0, \mu_0)$ is depicted as the black-filled circle, the MLE/MO is depicted by a circle, and the DL estimate is depicted by the diamond symbol.

6.6   Exact confidence inference for parameters of the beta distribution

The goal of this section is to illustrate the exact optimal statistical inference with the two shape parameters of the beta distribution. Unlike in the previous section, we use statistical simulations to approximate integrals. While this approach is time-consuming it is general and can be applied to any distribution. Several papers discuss asymptotic methods that work for large sample sizes (Marques et al. 2019), but the exact statistical inference for the shape parameters with a small sample is nonexistent, as far as we can tell. Here we apply the DL test and confidence region as they are well-suited for simulations.

We are given a random sample of size n from a beta distribution:

$$Y_i \overset{\text{iid}}{\sim} \mathcal{B}(\cdot; \alpha, \beta), \ i = 1, 2, ..., n, \tag{6.21}$$

where $\mathcal{B}$ denotes the beta distribution with the pdf given by

$$f(y; \sigma, \beta) = \frac{\Gamma(\alpha + \beta)}{\Gamma(\alpha)\Gamma(\beta)}y^{\alpha-1}(1 - y)^{\beta-1}, \quad 0 < y < 1$$

with positive shape parameters α and β of interest. The joint pdf is defined as the product

$$f(S_1, S_2; \alpha, \beta) = e^{nL(\alpha+\beta)-nL(\alpha)-nL(\beta)+(\alpha-1)S_1+(\beta-1)S_2}, \tag{6.22}$$

where

$$S_1 = \sum_{i=1}^{n} \ln Y_i, \quad S_2 = \sum_{i=1}^{n} \ln(1 - Y_i)$$

are sufficient statistics, and L denotes the log gamma function. We aim to develop the DL hypothesis test for the null hypothesis $H_0 : \alpha = \alpha_0$ and $\beta = \beta_0$.

Figure 6.3 displays the contours of the pdf (6.22) as a function of α and β given the observed sample $\mathbf{Y} = (Y_1, Y_2, ..., Y_n)$. The shape of these curves is the same as the DL confidence boundary to be discussed later. Note that the contours of the pdf have a somewhat elliptical shape along the 45° line. The filled circle shows the true α and β for which the sample $\mathbf{Y}$ was drawn. This plot was created by calling siBETA(job=0) with the following default values: n=10 is the sample size, alpha.true=2 is the true α, beta.true=3 is the true β, lambda=0.95 is the confidence level $\lambda = 1 - \alpha$, N=20 is the number of grid values, ss=4 is the seed number, nSim=10000 is the number of simulations, maxiter=100 is the number of maximum iterations, and eps=10^-6 is the tolerance limit for iterations.

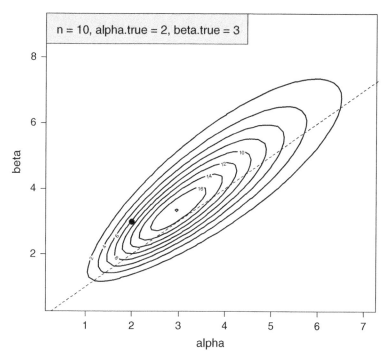

Figure 6.3: *Contours of the pdf (6.22) as a function of the shape parameters α and β given sample Y. The filled circle depicts the true value of α and β for which the sample Y was drawn. Note that the contours have a somewhat skewed elliptical shape along the 45° line. This plot was created by calling* siBETA(job=0).

6.6.1   Statistical tests

We are concerned with testing the null hypothesis $H_0 : \alpha = \alpha_0, \beta = \beta_0$ versus $H_A : (\alpha, \beta) \neq (\alpha_0, \beta_0)$.

First, we apply the traditional likelihood ratio test with the log-likelihood function given by

$$l(\alpha, \beta) = nL(\alpha + \beta) - nL(\alpha) - nL(\beta) + (\alpha - 1)S_1 + (\beta - 1)S_2.$$

The derivatives with respect to the unknown parameters are:

$$\frac{\partial l}{\partial \alpha} = n\psi(\alpha + \beta) - n\psi(\alpha) + S_1, \ \frac{\partial \ln l}{\partial \beta} = n\psi(\alpha + \beta) - n\psi(\beta) + S_2,$$

where ψ stands for the `digamma` function as the derivative of the log gamma function, $\psi = L' = (\ln \Gamma)'$. The 2×2 Fisher information matrix takes the form

$$\mathbf{H} = -n \begin{bmatrix} \eta(\alpha + \beta) - \eta(\alpha) & \eta(\alpha + \beta) \\ \eta(\alpha + \beta) & \eta(\alpha + \beta) - \eta(\beta) \end{bmatrix},$$

where η stands for the `trigamma` function as the derivative of ψ. The inverse of this matrix approximates the covariance matrix of the MLE $(\widehat{\alpha}, \widehat{\beta})$. According to the likelihood ratio test the null hypothesis is rejected if

$$2(l(\widehat{\alpha}, \widehat{\beta}) - l(\alpha_0, \beta_0)) > \chi_2^{-1}(\lambda), \tag{6.23}$$

where $\chi_2^{-1}(\lambda)$ is the λth quantile of the chi-square distribution with two df. The p-value is computed as the upper tail probability evaluated at D, the left-hand side of (6.23), the deviance.

Now we discuss how to test the null hypothesis with the DL test. According to this test the null hypothesis is rejected if $f(S_1, S_2; \alpha_0, \beta_0) < \kappa_0$, where κ_0 is found from the equation

$$\int_{\{y:f(s_1,s_2;\alpha_0,\beta_0)\geq\kappa_0\}} f(s_1, s_2; \alpha_0, \beta_0)dy_1 \cdots dy_n = \lambda, \tag{6.24}$$

where

$$s_1 = \sum_{i=1}^{n} \ln y_i, \quad s_2 = \sum_{i=1}^{n} \ln(1 - y_i).$$

The quasi-Newton's algorithm (6.2) applies where the maximum f is found from differentiation with respect to y_i at the null values:

$$M = f(S_1^*, S_2^*; \alpha_0, \beta_0),$$

$$S_1^* = \sum_{i=1}^{n} \ln y_i^* \quad S_2^* = \sum_{i=1}^{n} \ln(1 - y_i^*), \quad y_i^* = \frac{\alpha_0 - 1}{\alpha_0 + \beta_0 - 2}.$$

To approximate the integral (6.24) with simulations, we rewrite the left-hand side as

$$\Pr(f(S_1, S_2; \alpha_0, \beta_0) \geq \kappa_0)$$

where $Y_1, ..., Y_n$ are generated from the beta distribution with the null shape parameters α_0 and β_0. Therefore the integral is approximated as the proportion of samples for which $f(S_1, S_2; \alpha_0, \beta_0) \geq \kappa_0$. We found that for the beta distribution starting with $\kappa_0 = 0$ Newton's algorithm (6.3) yields satisfactory results, however, the number of simulations must be fairly large, say, nSim=100000.

Testing the null hypothesis and computing the p-value for the two tests is illustrated by calling siBETA(job=1) with the output shown here.

```
> siBETA(job=1)
S1=-8.511867, S2=-7.115281
pdf observed=8.220551, kappa=0.6040315
p-value ML=0.4581481, p-value DL=0.38051
Time difference of 0.9105809 secs
```

Since $\mathbf{Y}$ is generated from the null distribution we expect that the observed pdf value is greater than κ_0 is close to λ, and the p-value is close to 1. This is confirmed by the above output with the two p-values being fairly close.

6.6.2 Confidence regions

To the best of our knowledge, there are no methods for constructing exact confidence regions for the parameters of the beta distribution. Our goal is to develop the density level (DL) confidence region with an exact $100\lambda\%$ confidence level having a random sample $(Y_1, ..., Y_n)$ from the beta distribution with the shape parameters α and β as specified in (6.21). Although S_1 and S_2 are sufficient statistics we use the entire sample for our construction because the joint bivariate distribution of (S_1, S_2) is too complicated, so the general method of integral approximation via simulations in R^n is used. The DL confidence region is contrasted with two asymptotic methods: likelihood ratio (LR) and Wald ellipse confidence regions. As follows from (6.23), the LR confidence region for (α, β) consists of the points for which

$$2(l(\widehat{\alpha}, \widehat{\beta}) - l(\alpha, \beta)) \leq \chi_2^{-1}(\lambda).$$

To plot the boundary of this confidence region, short confidence curve, we compute the left-hand side on the grid of values for α and β and then use contour with levels equal to the right-hand side; see Figure 6.4 which was created by calling siBETA(job=3). The Wald confidence region consists of the points for which

$$\begin{bmatrix} \widehat{\alpha} - \alpha \\ \widehat{\beta} - \beta \end{bmatrix}' \mathbf{H} \begin{bmatrix} \widehat{\alpha} - \alpha \\ \widehat{\beta} - \beta \end{bmatrix} \leq \chi_2^{-1}(\lambda),$$

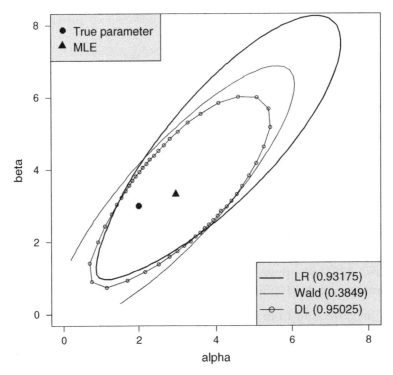

Figure 6.4: *The 95% confidence regions for three methods: likelihood-ratio, Wald, and DL with coverage probabilities in the parentheses with the true shape parameters $\alpha = 2$ and $\beta = 3$, $n = 10$.*

with the confidence curve as the ellipse at the center $(\widehat{\alpha}, \widehat{\beta})$. The method of displaying the curve is the same as in the previous method using the `contour` function. The local function `mle` finds $\widehat{\alpha}$ and $\widehat{\beta}$ and computes the matrix $\mathbf{H}$. The method of moments estimates

$$\widehat{\alpha}_{MM} = \overline{Y}\left(\frac{\overline{Y}(1-\overline{Y})}{\widehat{\sigma}^2} - 1\right), \quad \widehat{\beta}_{MM} = (1-\overline{Y})\left(\frac{\overline{Y}(1-\overline{Y})}{\widehat{\sigma}^2} - 1\right)$$

are used as the starting values for the Fisher scoring algorithm (Demidenko 2020, p. 500).

Now we turn our attention to the DL confidence region. The boundary of this region consists of the points (α, β) on the plane for which

$$\int_{\{\mathbf{y}: f(\mathbf{y};\alpha,\beta) \geq f(\mathbf{Y};\alpha,\beta)\}} f(\mathbf{y}; \alpha, \beta) dy_1 \cdots dy_n = \lambda \tag{6.25}$$

where $\mathbf{y} = (y_1, y_2, ..., y_n)$ and $\mathbf{Y} = (Y_1, Y_2, ..., Y_n)$ is the observed random sample. With two parameters, it is advantageous to plot the confidence curve in polar coordinates. Since the MLE maximizes f, the pair $(\widehat{\alpha}, \widehat{\beta})$ is an inner point of the confidence curve on the plane specified by (6.25). Define

$$\alpha = \widehat{\alpha} + \rho \cos \theta, \quad \beta = \widehat{\beta} + \rho \sin \theta$$

where $0 < \theta < 2\pi$, and ρ is the distance from the MLE to the boundary curve on the ray with angle θ. To plot the boundary of the $100\lambda\%$ confidence region we run θ on a grid of values from 0 to 2π and for each θ, solve equation (6.25) for ρ using the built-in function uniroot.

The three confidence curves are shown in Figure 6.4. The coverage probability for each method is shown in parentheses. These probabilities are estimated via simulations by running siBETA2_out=siBETA(job=2,nSim=20000), where siBETA2_out is the R array with three coverage probabilities. Both asymptotic confidence regions cover the true shape parameters $\alpha = 2$ and $\beta = 3$ with probabilities smaller than the nominal. The coverage probability of the Wald method is especially poor– the confidence ellipse even goes to the negative part of the plane. The DL method is practically exact and yet has minimal area.

6.7   Two-sample binomial probability

This section illustrates an application of M-statistics to a multivariate discrete distribution with statistical inference for the two-sample binomial distribution. Specifically, the subject of the section is the simultaneous statistical inference for the two binomial probabilities p_1 and p_2 having binomial outcomes m_1 and m_2 from independent Bernoulli samples of size n_1 and n_2, respectively. We are interested in (a) testing the null hypothesis

$$H_0 : p_1 = p_{10}, \; p_2 = p_{20} \tag{6.26}$$

versus the alternative that at least one equality does not hold, and (b) constructing the dual confidence region on the plane that covers the true p_1 and p_2, given m_1 and m_2 with the confidence level close to λ. Remember that for discrete distributions the coverage probability cannot be exactly equal to λ. The DL test and the confidence region are compared with the traditional asymptotic Wald test and approximate confidence region.

6.7.1   Hypothesis testing

Since the joint distribution of the binomial pair is the product of the probabilities of the individual outcomes the DL test reduces to finding $\kappa > 0$ such that

$$\min_{\kappa} \sum_{\{(i,j): p_{i1}p_{j2} \geq \kappa\}} p_{i1}p_{j2} = \lambda \tag{6.27}$$

where $i = 0, 1, ..., n_1$ and $j = 0, 1, ..., n_2$, and

$$p_{1i} = \binom{n_1}{i} p_{10}^i (1 - p_{10})^{n_1 - i}, \quad p_{2j} = \binom{n_2}{j} p_{20}^j (1 - p_{20})^{n_2 - j}.$$

We use min in (6.27) to comply with the general definition of the size of the test which requires that if the size of the test cannot be α for all true parameter values it must be equal at some point but smaller otherwise, as in the case of a discrete distribution (Zacks 1971; Lehmann and Romano 2005; Schervish 2012). Therefore if m_1 and m_2 are two binomial outcomes, we reject (6.26) if $p_{m_1} p_{m_2} < \kappa$ with the p-value

$$p = \sum_{\{(i,j) : p_{i1} p_{j2} < p_{m_1} p_{m_2}\}} p_{i1} p_{j2}. \tag{6.28}$$

Alternatively, instead of solving (6.27), one may compute p and reject the null if its value is smaller than a prespecified threshold, such as 0.05. As in the continuous case, a good approximation for κ is $M\alpha$, where $M = \max p_{1i} \times \max p_{2j}$.

Figure 6.5 illustrates the solution of equation (6.27) with $n_1 = 10, p_{10} = 0.2$, and $n_2 = 20, p_{20} = 0.6$. The exact $\kappa = 0.0037$ is depicted by the solid vertical line and an approximate $\kappa_0 = M\alpha = 0.0027$ is shown by the dotted vertical line – see code bpMC73.

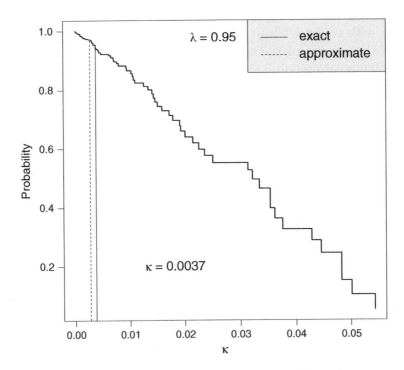

Figure 6.5: *Solution of (6.27) for $p_{10} = 0.2$ and $p_{02} = 0.6$ with two sample sizes $n_1 = 10$ and $n_2 = 20$. This graph was created by calling bpMC73(job=1).*

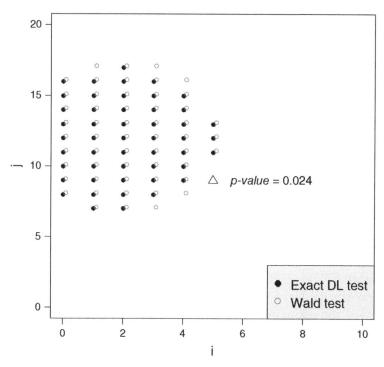

Figure 6.6: *Two acceptance regions for testing $p_{01} = 0.2$ and $p_{02} = 0.6$ with the sample sizes $n_1 = 10$ and $n_2 = 20$. This graph was created by calling* bpMC73(job=2)*.*

Figure 6.6 displays the acceptance region where the x-axis represents possible outcomes in the first sample $0, 1, ..., n_1$, and the y-axis represents possible outcomes in the second sample $0, 1, ..., n_2$. The triangle symbol depicts an example of a random outcome $(m_1 = 5, m_2 = 9)$. This outcome falls outside of the accepted region, shown with dots, and therefore the hypothesis that $p_1 = 0.2$ and $p_2 = 0.6$ is rejected with the p-value $= 0.024$ computed by formula (6.28). This figure was created by calling bpMC73(job=2). We make a few comments: (1) matrix P that contains values $\{p_{i1}p_{j2}\}$ is computed in a row-wise fashion using a single loop, (2) the values P are sorted in ascending order using the sort function, and (3) the left-hand side of (6.27) is computed as the sum of elements of matrix P greater than κ. The circles represent the Wald acceptance region computed as

$$\left\{ (i,j) : \frac{n_{10}(i/n_1 - p_{10})^2}{p_{10}(1 - p_{10})} + \frac{n_{20}(j/n_2 - p_{20})^2}{p_{20}(1 - p_{20})} \leq C_2^{-1}(\lambda) \right\}$$

where $C_2^{-1}(\lambda)$ is the λ quantile of the chi-square distribution with two df.

6.7.2 Confidence region

Let m_1 and m_2 be the observed number of successes from the two samples. The confidence region is dual to the DL test discussed earlier: it contains all pairs (p_1, p_2) on the unit square $(0, 1)^2$ for which the null hypothesis is accepted. First, on the grid of values for probability p_1 and p_2 of size N, we compute two $N \times N$ matrices. The entries of the first matrix $\kappa = \{\kappa_{ij}\}$ satisfy (6.27), and the second matrix $\mathbf{P}$ has the (i, j)th entry computed as

$$\binom{n_1}{m_1} p_{1i}^{m_1} (1 - p_{1i})^{n_1 - m_1} \times \binom{n_2}{m_2} p_{2j}^{m_2} (1 - p_{2j})^{n_2 - m_2}.$$

Second, the confidence region is the level set for which $\mathbf{P} - \kappa = 0$ (the `contour` function) is the boundary of the area $\{\mathbf{P} > \kappa\}$. The R code `bpMC73(job=3)` with $n_1 = 10, n_2 = 20$, and observed $m_1 = 2$ and $m_2 = 4$ produce the 95% confidence region in Figure 6.7.

Two approximate large-sample Wald confidence regions are suggested. The first region uses the estimates of the binomial proportions, called the *estimated Wald* method (Demidenko 2020; p. 574),

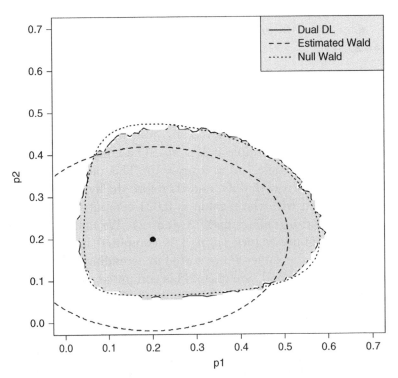

Figure 6.7: *Three 95% confidence regions for binomial probabilities with $m_1 = 2$ and $m_2 = 4$ of the sample sizes $n_1 = 10$ and $n_2 = 20$.*

$$\left\{ (p_1, p_2) : \frac{n_1(p_1 - \widehat{p}_1)^2}{\widehat{p}_1(1 - \widehat{p}_1)} + \frac{n_2(p_2 - \widehat{p}_2)^2}{\widehat{p}_2(1 - \widehat{p}_2)} \le q \right\}, \quad (6.29)$$

where $q = C_2^{-1}(1 - \alpha)$, the $(1 - \alpha)$ quantile of the chi-distribution with two df. This region defines an ellipse with axes parallel to the x and y-axes. The upper and lower arcs of the ellipse are found by solving a quadratic equation and are computed by the following formula:

$$p_2(p_1) = \widehat{p}_2 \pm \sqrt{n_2^{-1}\widehat{p}_2(1 - \widehat{p}_2)\left(q - \frac{n_1(p_1 - \widehat{p}_1)^2}{\widehat{p}_1(1 - \widehat{p}_1)} \right)},$$

where p_1 runs within the interval given by

$$\widehat{p}_1 - \sqrt{n_1^{-1}\widehat{p}_1(1 - \widehat{p}_1)q} \le p_1 \le \widehat{p}_1 + \sqrt{n_1^{-1}\widehat{p}_1(1 - \widehat{p}_1)q}.$$

A more sophisticated nonelliptical confidence region is defined when the variances in both samples are evaluated at the respective probability value, that is,

$$\left\{ (p_1, p_2) : \frac{n_1(p_1 - \widehat{p}_1)^2}{p_1(1 - p_1)} + \frac{n_2(p_2 - \widehat{p}_2)^2}{p_2(1 - p_2)} \le q \right\}. \quad (6.30)$$

This region is referred to as the *null Wald* method. To plot this region first we find the leftmost and rightmost values for p_1 by solving the quadratic equation $(x - \widehat{p}_1)^2/[x(1 - x)] = q/n_1$. The discriminant is $D_1 = q^2/n_1^2 + 4\widehat{p}_1(1 - \widehat{p}_1)q/n_1$, so that the area defined by (6.30) is contained on the x-axis between

$$x_{1,2} = \frac{2\widehat{p}_1 + q/n_1 \pm \sqrt{D_1}}{2(1 + q/n_1)}.$$

Second, for each $p_1 \in (x_1, x_2)$ we compute the upper and lower arcs of region (6.30),

$$p_2 = \frac{2\widehat{p}_2 + q_2 \pm \sqrt{D_2}}{2(1 + q_2)},$$

where

$$q_2 = \frac{1}{n_2}\left[q - \frac{n_1(p_1 - \widehat{p}_1)^2}{p_1(1 - p_1)} \right], \quad D_2 = q_2^2 + 4\widehat{p}_2(1 - \widehat{p}_2)q_2.$$

Unlike the estimated Wald test, this region does not have an elliptical shape.

Three CRs are shown in Figure 6.7 with observed $m_1 = 2$ and $m_2 = 4$ and $n_1 = 10$ and $n_2 = 20$, respectively. The filled circle depicts $\widehat{p}_1 = m_1/n_1 = 0.2$ and $\widehat{p}_2 = m_2/n_2 = 0.2$. The estimated Wald confidence region (dashed line) is an ellipse with the principal axes parallel to the x- and y-axes and covers negative p_1. The null Wald confidence region (dotted line) does not have an elliptic shape and is contained in the unit square. The grey area is where $\mathbf{P} > \kappa$ elementwise. The CR, dual to the DL test, has a jigsaw shape due to the discreteness of the binomial distribution. The null Wald and dual CRs are very close even for these fairly small sample sizes.

6.8   Exact and profile statistical inference for nonlinear regression

Nonlinear regression is a popular statistical model. Several special books, such as Bates and Watts (1988) and Seber and Wild (1989), and many papers are devoted to statistical inference based on the nonlinear least squares (NLS) estimator. Most of the methods rely on asymptotic properties, including high-order asymptotics (Brazalle and Davison 2008). However, only a handful of authors have worked on the moderate and small sample properties of the NLS estimator. The earliest attempt to study small-sample properties was by Hartley (1964). This work was followed by Pazman (2013, 2019) and Furlan and Mortarino (2020) and called a "near" exact statistical inference. Demidenko (2017) derived the exact density of the NLS estimator for a single parameter under the assumption that this estimator exists and is unique.

Unlike previous authors, we develop exact statistical inferences based on the pivotal statistic that works for small and extremely small sample sizes. Two types of statistical inference are discussed in this section: the exact statistical inference concerning the whole parameter vector and an approximate individual parameter of interest via profiling over the nuisance parameters. This section offers power functions for the developed tests as well. One of the goals of this section is to demonstrate how numerical issues of the nonlinear regression model, such as violation of the full rank or existence of several local minima of the residual sum of squares, complicate statistical inference, especially valuable when the sample size is small. As for many nonlinear statistical problems, studying small-sample properties is a formidable task because of the possibility of nonexistence and non-uniqueness of the NLS estimate (Demidenko 2000). For example, even for a single parameter, under mild assumptions, the residual sum of squares has several local minima with positive probability (Demidenko 2000). Needless to say, statistical inference for nonlinear regression is inseparable from the development of effective numerical solutions and algorithms. Our discussion is illustrated with exponential regression models.

The nonlinear regression is written in the vector form as

$$\mathbf{y} = \mathbf{f}(\boldsymbol{\beta}) + \varepsilon, \quad \varepsilon \sim \mathcal{N}(\mathbf{0}, \sigma^2 \mathbf{I}),$$

where $\mathbf{f} : R^m \to R^n$ is a vector-valued function of the m-dimensional parameter $\boldsymbol{\beta}$. It is assumed that $\boldsymbol{\beta}$ belongs to a fixed open set in R^m and that the regression is identifiable: $\mathbf{f}(\boldsymbol{\beta}_1) = \mathbf{f}(\boldsymbol{\beta}_2)$ implies $\boldsymbol{\beta}_1 = \boldsymbol{\beta}_2$. Define the $m \times 1$ vector and the $m \times m$ matrix as

$$\mathbf{g}(\boldsymbol{\beta}) = \mathbf{G}'(\boldsymbol{\beta})(\mathbf{y} - \mathbf{f}(\boldsymbol{\beta})), \quad \mathbf{H}(\boldsymbol{\beta}) = \mathbf{G}(\boldsymbol{\beta})(\mathbf{G}'(\boldsymbol{\beta})\mathbf{G}(\boldsymbol{\beta}))^{-1}\mathbf{G}'(\boldsymbol{\beta}),$$

where the $n \times m$ matrix $\mathbf{G}(\beta) = \partial \mathbf{f}/\partial \beta$ is the Jacobian. It is assumed that the Jacobian has a full rank for all β meaning the matrix $\mathbf{G}'(\beta)\mathbf{G}(\beta)$ is nonsingular. Define two quadratic forms as

$$Q_1(\beta) = (\mathbf{y} - \mathbf{f}(\beta))'\mathbf{H}(\beta)(\mathbf{y} - \mathbf{f}(\beta)),$$
$$Q_2(\beta) = (\mathbf{y} - \mathbf{f}(\beta))'(\mathbf{I} - \mathbf{H}(\beta))(\mathbf{y} - \mathbf{f}(\beta)). \qquad (6.31)$$

It is easy to see that for any fixed β, the normalized quadratic forms have chi-square distributions:

$$\sigma^{-2}Q_1(\beta) \sim \chi^2(m), \quad \sigma^{-2}Q_2(\beta) \sim \chi^2(n-m).$$

These forms are independent because matrix $\mathbf{H}(\beta)$ is idempotent and therefore $\mathbf{H}(\beta)(\mathbf{I} - \mathbf{H}(\beta)) = \mathbf{0}$ for every β (Rao 1973). Finally, define the pivotal quantity as

$$Q(\beta) = \frac{Q_1(\beta)/m}{Q_2(\beta)/(n-m)} \sim F(m, n-m), \qquad (6.32)$$

where F refers to the F-distribution with m and $n-m$ df. This pivot is the basis for our exact inference presented in the following section.

6.8.1 Statistical inference for the whole parameter

Confidence region

As follows from (6.32) the $100\lambda\%$ pivotal confidence region for the whole parameter β is defined as

$$\{\beta \in R^m : Q(\beta) \le q_\lambda\}, \qquad (6.33)$$

where $q_\lambda = q_\lambda(m, n-m)$ is the λth quantile of the F-distribution. When $\lambda \to 0$, the confidence region converges to a point, the NLS estimate as the solution of $Q(\beta) = 0$ or, equivalently, $\mathbf{G}'(\beta)(\mathbf{y} - \mathbf{f}(\beta)) = \mathbf{0}$.

Numerical complications, when studying small-sample statistical properties in nonlinear models, are illustrated in Figure 6.8. Two examples of the confidence region for the exponential regression

$$\mathbf{y} = \beta_1 e^{\beta_2 \mathbf{x}} + \varepsilon, \qquad (6.34)$$

where $\mathbf{x} = (-4, -1, 2, 3, 4, 5)'$ and $\sigma^2 = 0.5$, $\beta_1 = 1$, and $\beta_2 = 0.1$ are plotted as contours of the function $Q(\beta_1, \beta_2)$ defined by (6.32) with three confidence levels $\lambda = 95\%$, 80%, and 50%. The call to the R function that produces the figure is as follows:

```
nrEXPcr(sigma2=0.5,b1.true=1,b2.true=.1,
    x=c(-4,-1,2,3,4,5),N=200,ss=24)
```

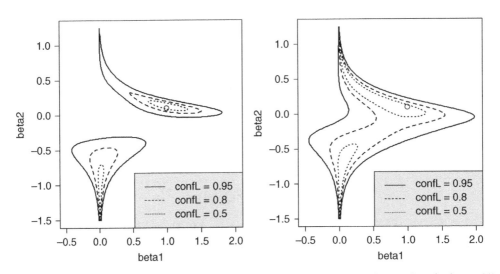

Figure 6.8: *Two examples of confidence regions with confidence levels λ = 95, 80, and 50% for nonlinear regression function $\beta_1 e^{\beta_2 x_i}$. The circle depicts the true parameter value. This graph was created by the function nrEXPcr.*

where sigma2 = σ^2, b1.true and b2.true specify the values for β_1 and β_2, the array x specifies values x_i, N is the number of points in the grid over β_1 and β_2 to compute the contours and ss is the seed number for simulations. The confidence regions are displayed by the built-in function contour with three levels corresponding to three values of λ. The two plots correspond to the two consecutive random simulations of y. This example illustrates that even a fairly simple nonlinear regression with a small sample size produces disconnected regions that indicate a possibility of several local minima of the residual sum of squares, resulting in the possibility of a false NLS estimate.

In the case of a single parameter ($m = 1$), the F-distribution in (6.32) can be replaced with the t-distribution with the lower and upper confidence limits for β derived by solving the equation

$$\frac{\mathbf{g}'(\beta)(\mathbf{y} - \mathbf{f}(\beta))\sqrt{n-1}}{\sqrt{\|\mathbf{y} - \mathbf{f}(\beta)\|^2 \|\mathbf{g}(\beta)\|^2 - (\mathbf{g}'(\beta)(\mathbf{y} - \mathbf{f}(\beta)))^2}} = \pm q_{1-\alpha/2}(n-1), \qquad (6.35)$$

where $\mathbf{g}(\beta) = d\mathbf{f}/d\beta$ is the $n \times 1$ vector. The t-distribution is more advantageous than the F-distribution because it may be used to construct an optimal unequal-tail CI shown shortly. Note that the left-hand side of (6.35) can be expressed as $r\sqrt{n-1}/\sqrt{1-r^2}$, where r is the cosine angle between vectors $\mathbf{g}(\beta)$ and $\mathbf{y} - \mathbf{f}(\beta)$.

Figure 6.9 shows the simulation-derived coverage probabilities of the traditional N-Wald/asymptotic normal 95% CI computed as $\widehat{\beta} \pm Z_{0.975}\widehat{\sigma}_\beta$, the T-Wald/asymptotic normal CI, which is the same but with the normal quantile

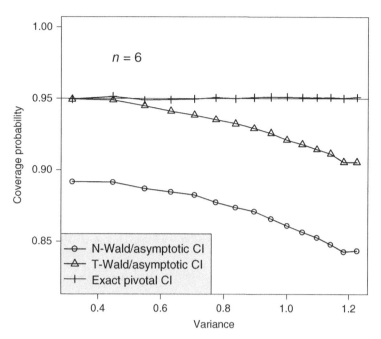

Figure 6.9: *The simulation-derived coverage probability for three 95% CIs for parameter β in the exponential regression $y_i = e^{\beta x_i} + \varepsilon_i$ (number of simulations = 100,000). The exact pivotal CI derived from (6.35) has coverage very close to nominal while the coverage of the traditional asymptotic normal CI is considerably smaller than 0.95 and decreases with σ^2. Using the t-distribution quantile improves the coverage but it goes down with σ^2 as well.*

replaced with t-distribution quantiles computed as $\widehat{\beta} \pm q_{0.975}(5) \times \widehat{\sigma}_\beta$, and the exact pivotal CI (6.35) as a function of σ^2 for an exponential regression without the coefficient β_1, that is, $\mathbf{y} = e^{\beta \mathbf{x}} + \varepsilon$.

A few comments on the code exp1SIM. It runs two jobs: the call

```
exp1SIM.out=exp1SIM(job=1,alpha=0.05,nSim=100000,ss=3)
```

performs simulations with $\alpha = 0.05$ and saves the coverage probabilities for each method in the file exp1SIM.out. The argument ss controls the seed parameter that helps to understand how the results vary depending on the number of simulations.

The call exp1SIM(job=2) plots Figure 6.9 using the data saved in the file exp1SIM.out. The package nleqslv is employed for solving equation (6.35) for β. Although it is tempting to start iterations from the N- or T-Wald confidence limits we found that more accurate starting values should be sought. In our code, we use the loop over the values of the confidence limit with the step 0.1 times the standard error of the parameter until the left-hand side is greater (or smaller) than the right-hand of (6.35). More efficient algorithms for finding the optimal starting values may be developed. As follows from this figure, the N-Wald/asymptotic normal CI severely undercovers the true parameter. The use of the t-quantile that accounts for the low df improves the coverage but deteriorates with the scatter around the regression. On the other hand, the coverage probability of the pivotal CI is very close to the nominal level on the entire range of σ^2. An advantage of the pivotal CI is that it is invariant with respect to a monotonic transformation of the parameter.

Now we outline the possibility of optimizing the CI by using unequal-tail probabilities. Specifically, we seek critical points of the t-distribution, $q_L < q_U$, such that $\mathcal{T}(q_U) - \mathcal{T}(q_L) = \lambda$ with minimum $\beta_U - \beta_L$, where $\mathcal{T}$ denotes the cdf of the t-distribution with $n-1$ df, and β_U and β_L are the solutions of (6.35) with the right-hand side q_L and q_U, respectively. One can replace this optimization problem with the system of two equations $\mathcal{T}(h(\beta_L)) - \mathcal{T}(h(\beta_U)) = \lambda$ and

$$t(h(\beta_L))h'(\beta_L) - t(h(\beta_U))h'(\beta_U) = 0,$$

where $t = \mathcal{T}'$, $h(\beta)$ is the left-hand side of (6.35), and $h'(\beta)$ is its derivative.

Hypothesis testing

Now we turn our attention to hypothesis testing. In accordance with (6.32) the null hypothesis $H_0 : \boldsymbol{\beta} = \boldsymbol{\beta}_0$ against $H_0 : \boldsymbol{\beta} \neq \boldsymbol{\beta}_0$ is rejected if

$$\frac{(\mathbf{y} - \mathbf{f}(\boldsymbol{\beta}_0))'\mathbf{H}(\boldsymbol{\beta}_0)(\mathbf{y} - \mathbf{f}(\boldsymbol{\beta}_0))/m}{(\mathbf{y} - \mathbf{f}(\boldsymbol{\beta}_0))'(\mathbf{I} - \mathbf{H}(\boldsymbol{\beta}_0))(\mathbf{y} - \mathbf{f}(\boldsymbol{\beta}_0))/(n-m)} > q_\lambda. \tag{6.36}$$

To derive the power function, we observe that under the alternative the numerator and denominator have independent noncentral chi-square distributions:

$$\sigma^{-2}Q_1(\boldsymbol{\beta}) \sim \chi^2(m, \mathrm{ncp} = \delta_1), \quad \sigma^{-2}Q_2(\boldsymbol{\beta}) \sim \chi^2(n - m, \mathrm{ncp} = \delta_2),$$

where the noncentrality parameters are given by

$$\delta_1 = \sigma^{-2}(\mathbf{f}(\boldsymbol{\beta}_0) - \mathbf{f}(\boldsymbol{\beta}))'\mathbf{H}(\boldsymbol{\beta}_0)(\mathbf{f}(\boldsymbol{\beta}_0) - \mathbf{f}(\boldsymbol{\beta})),$$
$$\delta_2 = \sigma^{-2}(\mathbf{f}(\boldsymbol{\beta}_0) - \mathbf{f}(\boldsymbol{\beta}))'(\mathbf{I} - \mathbf{H}(\boldsymbol{\beta}_0))(\mathbf{f}(\boldsymbol{\beta}_0) - \mathbf{f}(\boldsymbol{\beta})).$$

The random variable on the left-hand side of equation (6.36) follows a doubly noncentral F-distribution (Chattamvelli 1995) because in general $\delta_2 > 0$. However, for linear regression, $\mathbf{f}(\boldsymbol{\beta}) = \mathbf{X}\boldsymbol{\beta}$, we have $\delta_2 = 0$, and the distribution of the ratio turns into the traditional noncentral F-distribution (Demidenko 2020, p. 658). Although there is a special R package sadists for computing the cdf of a doubly noncentral F-distribution, we found that the approximations used there are unstable – here, we use the integral computation. Indeed, since the two distributions are independent, we can write the power function as the probability of rejection expressed via the integral

$$P(\beta, \beta_0) = 1 - \int_0^\infty C_m \left(\frac{q_\lambda m}{n-m} x, \mathrm{ncp} = \delta_1 \right) c_{n-m}(x, \mathrm{ncp} = \delta_2) dx,$$

where C and c stand for the cdf and pdf of the chi-square distribution with the specified df and ncp. This power function for a single parameter expressed via doubly noncentral F-distribution was derived earlier by Ittrich and Richter (2005). The test (6.36) is locally unbiased – the derivative of the power function with respect to β evaluated at the null value β_0 vanishes because, as it is easy to see,

$$\left. \frac{\partial \delta_1}{\partial \beta} \right|_{\beta=\beta_0} = \left. \frac{\partial \delta_2}{\partial \beta} \right|_{\beta=\beta_0} = \mathbf{0}.$$

Figure 6.10 depicts the contours of the power function for testing the null hypothesis $H_0 : \beta_1 = 1, \beta_2 = 0.1$ in the exponential nonlinear regression $y_i = \beta_1 e^{\beta_2 x_i} + \varepsilon_i$ with $\sigma = 0.5$ and x=c(-4,-1,2,3,4,5). Note that high power

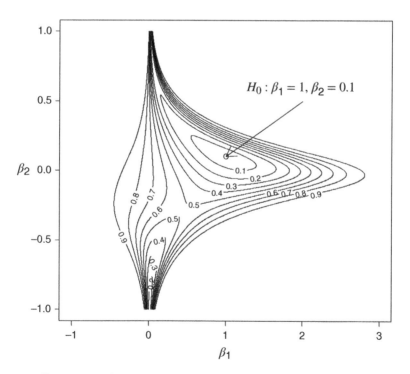

Figure 6.10: *Contours of the power function for testing $\beta_1 = 1, \beta_2 = 0.1$ in the exponential regression $y_i = \beta_1 e^{\beta_2 x_i} + \varepsilon_i$ with $\sigma = 0.5$. The high power may not be achieved for alternatives in the bottom left corner and several local minima are expected – typical complications of nonlinear regression.*

cannot be achieved for small alternative values and this regression may have several local minima – a frequently occurring situation in nonlinear regression with a small sample size. For this regression, $\beta_1 = 0$ nulls the regression function, and the power is β_2 independent: that is, it has a vertical asymptote.

6.8.2 Statistical inference for an individual parameter of interest via profiling

Often we are interested in an individual parameter of interest – without loss of generality we can assume that this is the first component of vector $\boldsymbol{\beta}$. Split $\boldsymbol{\beta} = (\beta, \boldsymbol{\gamma})$, where β is treated as the parameter of interest and $\boldsymbol{\gamma}$ is the $(m-1)$ dimensional nuisance parameter. It is important to realize that while the whole parameter inference discussed earlier is exact, profiling does not guarantee the exactness, although it may be very close to exact.

Following the line of the previous research popularized by Bates and Watts (1988), the nuisance parameters are eliminated via the profile function

$$\mathbf{G}'_{\boldsymbol{\gamma}}(\mathbf{y} - \mathbf{f}(\beta, \boldsymbol{\gamma})) = \mathbf{0}, \tag{6.37}$$

where $\mathbf{G}_{\boldsymbol{\gamma}} = \partial \mathbf{f} / \partial \boldsymbol{\gamma}$ is the $n \times (m-1)$ matrix of derivatives with respect to the nuisance parameter $\boldsymbol{\gamma}$. Equivalently, $\boldsymbol{\gamma}$ can be derived as the minimizer of the residual sum of squares, $\|\mathbf{y} - \mathbf{f}(\beta, \boldsymbol{\gamma})\|^2$, when β is held fixed. Consequently, one can use regular software, such as the R function nls, to get $\widehat{\boldsymbol{\gamma}}$ for each β. The solution of the system of equations (6.37) enables us to define $\widehat{\boldsymbol{\gamma}} = \widehat{\boldsymbol{\gamma}}(\beta)$; therefore, the individual CI for β is derived from (6.35) but with $\boldsymbol{\gamma}$ replaced with $\widehat{\boldsymbol{\gamma}}(\beta)$, which leads to

$$\frac{\mathbf{g}'(\beta)(\mathbf{y} - \mathbf{f}(\beta, \widehat{\boldsymbol{\gamma}}(\beta)))\sqrt{n-m}}{\sqrt{\|\mathbf{y} - \mathbf{f}(\beta, \widehat{\boldsymbol{\gamma}}(\beta))\|^2 \|\mathbf{g}(\beta)\|^2 - (\mathbf{g}'(\beta)(\mathbf{y} - \mathbf{f}(\beta, \widehat{\boldsymbol{\gamma}}(\beta))))^2}} = \pm q_{1-\alpha/2}(n-m).$$
$$\tag{6.38}$$

However, before solving this equation we need to find the profile derivative

$$\mathbf{g}(\beta) = \frac{d\mathbf{f}(\beta, \widehat{\boldsymbol{\gamma}}(\beta))}{d\beta} = \frac{d\mathbf{f}}{d\beta} + \frac{\partial \mathbf{f}}{\partial \widehat{\boldsymbol{\gamma}}} \frac{d\widehat{\boldsymbol{\gamma}}}{d\beta} = \mathbf{g}_\beta + \mathbf{G}_{\boldsymbol{\gamma}} \frac{d\widehat{\boldsymbol{\gamma}}}{d\beta},$$

where $\mathbf{g}_\beta = d\mathbf{f}(\beta, \boldsymbol{\gamma})/d\beta$. To find $d\widehat{\boldsymbol{\gamma}}/d\beta$ we use the formula for the derivative of the implicitly defined function $\widehat{\boldsymbol{\gamma}} = \widehat{\boldsymbol{\gamma}}(\beta)$ as the solution of the system (6.37). Define the following $(m-1) \times n(m-1)$ and $(m-1) \times n$ matrices:

$$\mathbf{K} = \left[\frac{\partial^2 f_1}{\partial^2 \boldsymbol{\gamma}}, \frac{\partial^2 f_2}{\partial^2 \boldsymbol{\gamma}}, \dots, \frac{\partial^2 f_n}{\partial^2 \boldsymbol{\gamma}}\right], \quad \mathbf{J} = \left[\frac{\partial^2 f_1}{\partial \boldsymbol{\gamma} \partial \beta}, \frac{\partial^2 f_2}{\partial \boldsymbol{\gamma} \partial \beta}, \dots, \frac{\partial^2 f_n}{\partial \boldsymbol{\gamma} \partial \beta}\right].$$

Then in the matrix language, we obtain

$$\frac{d\widehat{\boldsymbol{\gamma}}}{d\beta} = -\left(\mathbf{G}'_{\boldsymbol{\gamma}}\mathbf{G}_{\boldsymbol{\gamma}} - \mathbf{K}(\mathbf{y} - \mathbf{f}) \otimes \mathbf{I}_{m-1}\right)^{-1}\left(\mathbf{G}'_{\boldsymbol{\gamma}}\mathbf{g}_\beta - \mathbf{J}(\mathbf{y} - \mathbf{f})\right),$$

where $\otimes$ denotes the matrix Kronecker product. Finally, the profile derivative is given by

$$\mathbf{g}(\beta) = \mathbf{g}_\beta - \mathbf{G}_\gamma \left(\mathbf{G}'_\gamma \mathbf{G}_\gamma - \mathbf{K}(\mathbf{y} - \mathbf{f}) \otimes \mathbf{I}_{m-1}\right)^{-1} \left(\mathbf{G}'_\gamma \mathbf{g}_\beta - \mathbf{J}(\mathbf{y} - \mathbf{f})\right). \quad (6.39)$$

Note that the matrix $\mathbf{G}'_\gamma \mathbf{G}_\gamma - \mathbf{K}(\mathbf{y} - \mathbf{f}) \otimes \mathbf{I}_{m-1}$ is positive definite under mild conditions. Indeed, when β is fixed the Hessian of the residual sum of squares can be written as

$$\frac{1}{2}\frac{\partial^2}{\partial \gamma^2}\|\mathbf{y} - \mathbf{f}(\beta, \gamma)\|^2 = \sum_{i=1}^n \left[\left(\frac{\partial f_i}{\partial \gamma}\right)\left(\frac{\partial f_i}{\partial \gamma}\right)' - (y_i - f_i(\beta, \gamma))\frac{\partial^2 f_i}{\partial \gamma^2} \right].$$

Under the condition(6.37), it is positive definite for every β as the Hessian matrix of the minimized function at the minimum of the residual sum of squares. This general approach is illustrated by some special cases.

Proposition 6.9 *The CI for β as the solution of (6.38) produces the traditional CI in linear model.*

Proof. Write a linear model as $\mathbf{y} = \beta\mathbf{z} + \mathbf{X}\gamma + \varepsilon$, where matrix $\mathbf{U} = [\mathbf{z}; \mathbf{X}]$ has full rank m. Then $\mathbf{g}_\beta = \mathbf{z}$, $\mathbf{G}_\gamma = \mathbf{X}$, $\mathbf{K} = \mathbf{0}$, and $\mathbf{J} = \mathbf{0}$. Equation (6.37) turns into $\mathbf{X}'(\mathbf{y} - \beta\mathbf{z} - \mathbf{X}\gamma) = \mathbf{0}$, which yields $\widehat{\gamma} = (\mathbf{X}'\mathbf{X})^{-1}\mathbf{X}'(\mathbf{y} - \beta\mathbf{z})$ and $d\widehat{\gamma}/d\beta = -(\mathbf{X}'\mathbf{X})^{-1}\mathbf{X}'\mathbf{z}$. Define $\mathbf{A} = \mathbf{I} - \mathbf{X}(\mathbf{X}'\mathbf{X})^{-1}\mathbf{X}'$. Then for the numerator and denominator we have

$$\mathbf{g}'(\beta)(\mathbf{y} - \mathbf{f}(\beta, \widehat{\gamma}(\beta))) = \widetilde{\mathbf{z}}'(\widetilde{\mathbf{y}} - \beta\widetilde{\mathbf{z}}),$$

$$\|\mathbf{y} - \mathbf{f}(\beta, \widehat{\gamma}(\beta))\|^2 \|\mathbf{g}(\beta)\|^2 - (\mathbf{g}'(\beta)(\mathbf{y} - \mathbf{f}(\beta, \widehat{\gamma}(\beta))))^2$$
$$= \|\widetilde{\mathbf{y}} - \beta\widetilde{\mathbf{z}}\|^2 \|\widetilde{\mathbf{z}}\|^2 - (\widetilde{\mathbf{z}}'(\widetilde{\mathbf{y}} - \beta\widetilde{\mathbf{z}}))^2$$

respectively, where $\widetilde{\mathbf{y}} = \mathbf{A}\mathbf{y}$ and $\widetilde{\mathbf{z}} = \mathbf{A}\mathbf{z}$. Now we use the following fact about linear models, sometimes referred to as the partial least squares:

$$\min_{\beta, \gamma} \|\mathbf{y} - \beta\mathbf{z} - \mathbf{X}\gamma\|^2 = \min_\beta \|\widetilde{\mathbf{y}} - \beta\widetilde{\mathbf{z}}\|^2.$$

This means the square root denominator in (6.37) divided by $\sqrt{n-m}$ is $\widehat{\sigma}$, and the equation turns into $(\mathbf{y} - \beta\mathbf{z})'\mathbf{A}\mathbf{z}/\sqrt{\mathbf{z}'\mathbf{A}\mathbf{z}} = \widehat{\sigma}q_{n-m}$. Finally, the upper and lower limits are

$$\frac{\mathbf{y}'\mathbf{A}\mathbf{z} \pm \widehat{\sigma}q_{n-m}\sqrt{\mathbf{z}'\mathbf{A}\mathbf{z}}}{\mathbf{z}'\mathbf{A}\mathbf{z}} = \widehat{\beta} \pm \widehat{\sigma}q_{n-m}\frac{1}{\sqrt{\mathbf{z}'\mathbf{A}\mathbf{z}}}.$$

But it's easy to see that $\sqrt{(\mathbf{U}'\mathbf{U})_{11}^{-1}} = 1/\sqrt{\mathbf{z}'\mathbf{A}\mathbf{z}}$ completes the proof. $\quad\square$

Proposition 6.10 *Let the regression function be defined as* $\mathbf{f}(\beta, \boldsymbol{\gamma}) = \mathbf{f}_0(\beta) + \mathbf{X}\boldsymbol{\gamma}$*, where* $\mathbf{X}$ *is the* $n \times (m-1)$ *fixed matrix of full rank. Then the pivotal equal-tail confidence limits for* β *are derived by solving the following equations:*

$$\frac{\widetilde{\mathbf{g}}'(\beta)(\widetilde{\mathbf{y}} - \widetilde{\mathbf{f}}(\beta))\sqrt{n-m}}{\sqrt{\left\|\widetilde{\mathbf{y}} - \widetilde{\mathbf{f}}(\beta)\right\|^2 \|\widetilde{\mathbf{g}}(\beta)\|^2 - (\widetilde{\mathbf{g}}'(\beta)(\widetilde{\mathbf{y}} - \widetilde{\mathbf{f}}(\beta)))^2}} = \pm q_{1-\alpha/2}(n-m),$$

where $\widetilde{\mathbf{y}} = \mathbf{A}\mathbf{y}$*,* $\widetilde{\mathbf{f}} = \mathbf{A}\mathbf{f}$*,* $\widetilde{\mathbf{g}} = \mathbf{A}(d\mathbf{f}_0/d\beta)$*, and* $\mathbf{A} = \mathbf{I} - \mathbf{X}(\mathbf{X}'\mathbf{X})^{-1}\mathbf{X}'$ *is the annihilator matrix.*

Proof. In the notation of (6.39), $\mathbf{G}_{\boldsymbol{\gamma}} = \mathbf{X}$ and $\widehat{\boldsymbol{\gamma}} = (\mathbf{X}'\mathbf{X})^{-1}\mathbf{X}'(\mathbf{y} - \mathbf{f}_0(\beta))$. Therefore, $d\widehat{\boldsymbol{\gamma}}/d\beta = -(\mathbf{X}'\mathbf{X})^{-1}\mathbf{X}'\mathbf{g}_{\beta}$ where $\mathbf{g}_{\beta} = d\mathbf{f}_0/d\beta$. It is easy to see that one arrives at the same derivative applying the general formula (6.39). Thus,

$$\mathbf{g}(\beta) = \mathbf{g}_{\beta} + \mathbf{G}_{\boldsymbol{\gamma}}\frac{d\boldsymbol{\gamma}}{d\beta} = \mathbf{g}_{\beta} - \mathbf{X}(\mathbf{X}'\mathbf{X})^{-1}\mathbf{X}'\mathbf{g}_{\beta} = \mathbf{A}\mathbf{g}_{\beta}.$$

The rest of the proof is obvious.   □

Proposition 6.11 *Let the two-parameter nonlinear regression function be defined as* $\mathbf{f}(\beta, \boldsymbol{\gamma}) = \boldsymbol{\gamma}\mathbf{f}_0(\beta)$*. The pivotal equal-tail lower and upper confidence limits for* β *are obtained as solutions of the following equation (the argument* β *in parentheses is omitted for brevity)*

$$\frac{\mathbf{g}'(\mathbf{y} - \widehat{\boldsymbol{\gamma}}\mathbf{f}_0)\sqrt{n-2}}{\sqrt{\|\mathbf{y} - \widehat{\boldsymbol{\gamma}}\mathbf{f}_0\|^2 \|\mathbf{g}\|^2 - (\mathbf{g}'(\mathbf{y} - \widehat{\boldsymbol{\gamma}}\mathbf{f}_0))^2}} = \pm q_{1-\alpha/2}(n-2),$$

where

$$\widehat{\boldsymbol{\gamma}} = \frac{\mathbf{y}'\mathbf{f}_0}{\|\mathbf{f}_0\|^2}, \quad \mathbf{g} = \widehat{\boldsymbol{\gamma}}\mathbf{g}_0 + \frac{\mathbf{y}'\mathbf{g}_0}{\|\mathbf{f}_0\|^2}\mathbf{f}_0 - 2\widehat{\boldsymbol{\gamma}}\frac{(\mathbf{g}_0'\mathbf{f}_0)}{\|\mathbf{f}_0\|^2}\mathbf{f}_0,$$

and $\mathbf{g}_0 = d\mathbf{f}_0/d\beta$*.*

Proof. Since $\boldsymbol{\gamma}$ is a linear parameter the solution of (6.37) is written in a closed form as $\widehat{\boldsymbol{\gamma}} = \mathbf{y}'\mathbf{f}_0/\|\mathbf{f}_0\|^2$. We have $\mathbf{g}_{\beta} = \widehat{\boldsymbol{\gamma}}\mathbf{g}_0$ and $\mathbf{G}_{\beta} = \mathbf{f}_0, \mathbf{K} = 0, \mathbf{J} = \mathbf{g}_0'$. Then using (6.39), we obtain

$$\mathbf{g} = \widehat{\boldsymbol{\gamma}}\mathbf{g}_0 - \frac{\widehat{\boldsymbol{\gamma}}\mathbf{g}_0'\mathbf{f}_0 - (\mathbf{y} - \widehat{\boldsymbol{\gamma}}\mathbf{f}_0)'\mathbf{g}_0}{\|\mathbf{f}_0\|^2}\mathbf{f}_0 = \widehat{\boldsymbol{\gamma}}\mathbf{g}_0 - \frac{\widehat{\boldsymbol{\gamma}}\mathbf{g}_0'\mathbf{f}_0 - \mathbf{y}'\mathbf{g}_0 + \widehat{\boldsymbol{\gamma}}\mathbf{f}_0'\mathbf{g}_0}{\|\mathbf{f}_0\|^2}\mathbf{f}_0$$

$$= \widehat{\boldsymbol{\gamma}}\mathbf{g}_0 - \frac{2\widehat{\boldsymbol{\gamma}}\mathbf{g}_0'\mathbf{f}_0 - \mathbf{y}'\mathbf{g}_0}{\|\mathbf{f}_0\|^2}\mathbf{f}_0,$$

the same as in the proposition statement.   □

Note that the previous CIs can be turned into optimal unequal-tail CIs for testing the double-sided hypothesis as discussed in Section 6.8.1.

Hypothesis testing for individual parameter

We aim at testing the double-sided null hypothesis $H_0 : \beta = \beta_0$ versus $H_A : \beta \neq \beta_0$ using the profile pivotal approach where the nonlinear regression function is $\mathbf{f}(\beta, \boldsymbol{\gamma})$. Here β is treated as the parameter of interest and $\boldsymbol{\gamma}$ as a nuisance parameter. The pivotal equal-tail test of type I error α rejects the null hypothesis if

$$\left| \frac{\mathbf{g}'(\beta_0)(\mathbf{y} - \mathbf{f}(\beta_0, \widehat{\boldsymbol{\gamma}}_0))\sqrt{n - m}}{\sqrt{\|\mathbf{y} - \mathbf{f}(\beta_0, \widehat{\boldsymbol{\gamma}}_0)\|^2 \|\mathbf{g}(\beta_0)\|^2 - (\mathbf{g}'(\beta_0)(\mathbf{y} - \mathbf{f}(\beta_0, \widehat{\boldsymbol{\gamma}}_0)))^2}} \right| > q_{1-\alpha/2}(n - m),$$

where $\widehat{\boldsymbol{\gamma}}_0$ is the solution to $\mathbf{G}'_{\boldsymbol{\gamma}}(\mathbf{y} - \mathbf{f}(\beta_0, \widehat{\boldsymbol{\gamma}}_0)) = \mathbf{0}$ and $\mathbf{g}(\beta_0)$ is computed by (6.39).

Now we derive the analytic power of the test. Besides the null β_0 and alternative value β, the power function depends on σ^2 and the $(m - 1)$-dimensional nuisance parameter $\boldsymbol{\gamma}$. Define the noncentrality parameters as follows:

$$\delta_1 = \frac{\mathbf{g}'(\beta_0)(\mathbf{f}(\beta, \boldsymbol{\gamma}) - \mathbf{f}(\beta_0, \widehat{\boldsymbol{\gamma}}_0))}{\sigma \|\mathbf{g}(\beta_0)\|},$$

$$\delta_2 = \frac{\|\mathbf{f}(\beta, \boldsymbol{\gamma}) - \mathbf{f}(\beta_0, \widehat{\boldsymbol{\gamma}}_0)\|^2 \|\mathbf{g}(\beta_0)\|^2 - (\mathbf{g}'(\beta_0)(\mathbf{f}(\beta, \boldsymbol{\gamma}) - \mathbf{f}(\beta_0, \widehat{\boldsymbol{\gamma}}_0)))^2}{\sigma^2 \|\mathbf{g}(\beta_0)\|^2},$$

where the value $\widehat{\boldsymbol{\gamma}}_0$ is obtained from the solution of the system of $m - 1$ equations

$$\mathbf{G}'_{\boldsymbol{\gamma}}(\mathbf{f}(\beta, \boldsymbol{\gamma}) - \mathbf{f}(\beta_0, \widehat{\boldsymbol{\gamma}}_0)) = \mathbf{0}.$$

The derivative $\mathbf{g}(\beta) = d\mathbf{f}(\beta, \widehat{\boldsymbol{\gamma}}(\beta))/d\beta$ is obtained via the implicitly defined function using formula (6.39) with $\mathbf{y}$ replaced by $\mathbf{f}(\beta, \boldsymbol{\gamma})$. Similarly to testing the whole parameter the power function reduces to the doubly noncentral t-distribution. Special statistical packages could be used to compute the required cdf but we found that expressing it via a one-dimensional integral is more computationally stable. Let the quantile $q_1 < q_2$ be such that $\mathcal{T}_{n-m}(q_2) - \mathcal{T}_{n-m}(q_1) = 1 - \alpha$. We accept the null if $q_1 < U\sqrt{n - m}/\sqrt{V} < q_2$, where $U \sim \mathcal{N}(\delta_1, 1)$ and $V \sim \chi^2(n - m, \mathrm{ncp} = \delta_2)$, distributed independently. Define

$$s(q) = \Pr\left(\frac{U\sqrt{n - m}}{\sqrt{V}} < q \right) = \Pr\left(U < \frac{q}{\sqrt{n - m}}\sqrt{V} \right)$$

$$= \int_0^\infty \left(\int_{-\infty}^{\frac{q}{\sqrt{n-m}}\sqrt{v}} f_U(u)du \right) f_V(v)dv$$

$$= \int_0^\infty \Phi\left(\frac{q}{\sqrt{n - m}}\sqrt{v} - \delta_1 \right) c_{n-m}(v, \mathrm{ncp} = \delta_2)dv.$$

Finally, the power of the profile pivotal test is given by

$$P(\beta; \beta_0, \boldsymbol{\gamma}) = 1 - s(q_2) + s(q_1).$$

The derivative of the power function at the null value is zero and the test is unbiased.

Following the line of proof in Proposition 6.9, one can show that for the linear model $\mathbf{y} = \mathbf{U}\boldsymbol{\beta} + \varepsilon$, where $\boldsymbol{\beta} = (\beta, \boldsymbol{\gamma})'$ and $\mathbf{U} = [\mathbf{z}; \mathbf{X}]$, the power does not depend on the nuisance parameter $\boldsymbol{\gamma}$ and can be expressed via the cdf of the noncentral t-distribution as

$$P(\beta; \beta_0) = 1 - \mathcal{T}_{n-m}(q_2; \mathrm{ncp} = \delta) + \mathcal{T}_{n-m}(q_1; \mathrm{ncp} = \delta),$$

where $\delta = \sigma^{-1}(\beta - \beta_0)/\sqrt{(\mathbf{U}'\mathbf{U})_{11}^{-1}}$ is the noncentrality parameter.

Example 6.12 *(a) Compare the simulation-derived coverage probability of the traditional asymptotic CI with the profile CI from (6.38) for the exponential regression $\mathbf{f}(\beta, \gamma) = \gamma e^{\beta \mathbf{x}}$. (b) Derive the analytic power function for testing $H_0 : \beta = \beta_0$ against $H_A : \beta_0 \neq \beta$ and check with simulations.*

Solution. (a) First, given β, γ, σ^2, and β_0, define $\widehat{\gamma}_0$ as the least squares estimate of the slope in the regression of $\gamma e^{\beta \mathbf{x}}$ on $\gamma_0 e^{\beta_0 \mathbf{x}}$: that is,

$$\widehat{\gamma}_0 = \gamma \frac{\sum e^{(\beta_0 + \beta) x_i}}{\sum e^{2\beta_0 x_i}} = \gamma \frac{\mathbf{1}' e^{(\beta_0 + \beta)\mathbf{x}}}{\mathbf{1}' e^{2\beta_0 \mathbf{x}}}.$$

Second, obtain the derivative of $\widehat{\gamma}_0$ with respect to β_0 as

$$\frac{d\widehat{\gamma}_0}{d\beta_0} = \gamma \left(\frac{\mathbf{x}' e^{(\beta_0 + \beta)\mathbf{x}}}{\mathbf{1}' e^{2\beta_0 \mathbf{x}}} - 2 \frac{\mathbf{1}' e^{(\beta_0 + \beta)\mathbf{x}}}{(\mathbf{1}' e^{2\beta_0 \mathbf{x}})^2} (\mathbf{x}' e^{2\beta_0 \mathbf{x}}) \right).$$

Third, derive

$$\mathbf{g}(\beta_0) = \frac{d}{d\beta_0} (\widehat{\gamma}_0 e^{\beta_0 \mathbf{x}}) = \frac{d\widehat{\gamma}_0}{d\beta_0} e^{\beta_0 \mathbf{x}} + \widehat{\gamma}_0 (\mathbf{x} e^{\beta_0 \mathbf{x}}).$$

Note that we arrive at the same expression using formula (6.39) formally letting $\mathbf{g}_\beta = \widehat{\gamma}_0 \mathbf{x} e^{\beta_0 \mathbf{x}}$, $\mathbf{K} = \mathbf{0}$, $\mathbf{J} = \mathbf{x} e^{\beta_0 \mathbf{x}}$, $\mathbf{y} = \gamma e^{\beta \mathbf{x}}$, $\mathbf{f} = \widehat{\gamma}_0 e^{\beta_0 \mathbf{x}}$ and $\mathbf{G}_\gamma = e^{\beta_0 \mathbf{x}}$. Fourth, plugging these expressions into formula (6.39) we obtain

$$\widehat{\gamma}_0 \mathbf{x} e^{\beta_0 \mathbf{x}} - \frac{\widehat{\gamma}_0 \mathbf{x}' e^{2\beta_0 \mathbf{x}} - (\mathbf{x} e^{\beta_0 \mathbf{x}})'(\gamma e^{\beta \mathbf{x}} - \widehat{\gamma}_0 e^{\beta_0 \mathbf{x}})}{\mathbf{1}' e^{2\beta_0 \mathbf{x}}} e^{\beta_0 \mathbf{x}},$$

which is equal to $\mathbf{g}(\beta_0)$ obtained earlier. (b) The previous quantities are plugged into expressions for the noncentrality parameters δ_1 and δ_2 to calculate s for $q = \pm q_{1-0.5/2}(n-2)$ and finally the power function $P(\beta; \beta_0, \gamma)$.

Now we provide some details of the code. Similarly to the previous calculations of the CI depicted in Figure 6.9 `job=1` performs simulations and `job=2` plots Figure 6.11. The simulation results are saved in the file `exp2SIM_out` upon issuing the command

```
exp2SIM_out=exp2SIM(job=1,alpha=0.05,nSim=10000,ss=3)
```

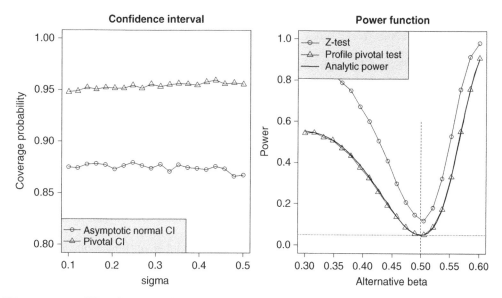

Figure 6.11: *The CIs and the power functions for β in the exponential regression $\gamma e^{\beta x}$. The circle and triangle values are derived from simulations (number of simulations = 50,000) and the smooth curve is computed from analytic power based on the doubly noncentral t-distribution.*

The output file is a list of two matrices. The first matrix contains the coverage probabilities for asymptotic and pivotal CIs, and the second matrix contains the corresponding power values. This job takes about 15 minutes. This code uses a different algorithm for solving equation (6.38). First, we employ the R package `numDeriv` and function `grad` to numerically approximate the derivative of the left-hand side of equation (6.38) with respect to β, and second, we use Newton's algorithm to find the solution.

Figure 6.11 depicts the 95% simulation-derived CI with the true $\beta_0 = 0.5$ and $\gamma = 1$. The asymptotic normal CI consistently undercovers the true β_0. The respective test has the type I error twice as large as the nominal 0.05, while the profile pivotal test has the exact significance level. Note that although the power function of the normal test is above the profile pivotal test the powers are incomparable because they have different type I errors.

References

[1] Ardia, D., Boudt, K., Carl, P., et al. (2011). Differential evolution with DEoptim. An application to non-convex portfolio optimization. *The R Journal* 3: 27–34.

[2] Arnold, B.C. and Shavelle, R.M. (1998). Joint confidence sets for the mean and variance of a normal distribution. *The American Statistician* 52: 133–140.

[3] Bates, D.M. and Watts, D.G. (1988). *Nonlinear Regression Analysis and Its Applications*. New York: Wiley.

[4] Brazzale, A.R. and Davison, A.C. (2008). Accurate parametric inference for small samples. *Statistical Science* 23: 465–484.

[5] Chattamvelli, R. (1995). On the doubly noncentral F distribution. *Computational Statistics & Data Analysis* 20: 481–489.

[6] Choudhari, P., Kundu, D., and Misra, N. (2001). Likelihood ratio test for simultaneous testing of the mean and the variance of a normal distribution. *Journal of Statistical Computation and Simulation* 71: 313–333.

[7] Demidenko, E. (2000). Is this the least squares estimate? *Biometrika* 87: 437–452.

[8] Demidenko, E. (2008). Criteria for unconstrained global optimization. *Journal of Optimization Theory and Applications* 136: 375–395.

[9] Demidenko E. (2017). Exact and approximate statistical inference for nonlinear regression and the estimating equation approach. *Scandinavian Journal of Statistics* 44: 636–665.

[10] Demidenko, E. (2020). *Advanced Statistics with Applications in R*. Hoboken, NJ: Wiley.

[11] Frey, J., Marrero, O., and Norton, D. (2009). Minimum-area confidence sets for a normal distribution. *Journal of Statistical Planning and Inference* 139: 1023–1032.

[12] Furlan, C. and Mortarino, C. (2020).Comparison among simultaneous confidence regions for nonlinear diffusion models. *Computational Statistics.* https://doi.org/10.1007/s00180-019-00949-0.

[13] Hartley, H.O. (1964). Exact confidence regions for the parameters in non-linear regression laws. *Biometrika* 51: 347–353.

[14] Ittrich, C. and Richter, W.-D. (2005). Exact tests and confidence regions in nonlinear regression. *Statistics* 39: 13–42.

[15] Lehmann, E.L. (1993). The Fisher, Neyman-Pearson theories of testing hypotheses: One Theory or Two? *Journal of the American Statistical Association* 88: 1242–1249.

[16] Lehmann, E.L. and Romano, J.P. (2005). *Testing Statistical Hypothesis*, 3e. New York: Springer.

[17] Marques, F.J., Coolen, F.P.A., and Coolen-Maturi, T. (2019). Approximation for the likelihood ratio statistic for hypothesis testing between two beta distributions. *Journal of Statistical Theory and Practice* 13: 17.

[18] Mood, A.M. (1950). *Introduction to the Theory of Statistics*. New York: McGraw-Hill.

[19] Neyman, J. and Pearson, E.S. (1936). Contributions to the theory of testing statistical hypotheses I. *Statistical Research Memoirs* 1: 1–37.

[20] Ortega, J.M. and Rheinboldt, W.C. (2000). *Iterative Solution of Nonlinear Equations in Several Variables*. Philadelphia: SIAM.

[21] Pazman, A. (2013). *Nonlinear Statistical Models*. Springer Science & Business Media.

[22] Pazman, A. (2019). Distribution of the multivariate nonlinear LS estimator under an uncertain input. *Statistical Papers* 60: 529–544.

[23] Pfeffer, W.F. (1987). The multidimensional theorem of calculus. *Journal of the Australian Mathematical Society, Ser. A* 43: 143–170.

[24] Rao, C.R. (1973). *Linear Statistical Inference and Its Applications*, 2e. New York: Wiley.

[25] Schervish, M. (2012). *Theory of Statistics*. New York: Springer.

[26] Seber, G.A.F. and Wild, C.J. (1989). *Nonlinear Regression*. New York: Wiley.

[27] Spivak, M. (2018). *Calculus on Manifolds: A Modern Approach to Classical Theorems of Advanced Calculus*. Boca Raton: CRC Press.

[28] Wilks, S.S. (1962). *Mathematical Statistics*. New York: Wiley.

[29] Zacks, S. (1971). *The Theory of Statistical Inference*. New York: Wiley.

[30] Zhang, J. (2017). Minimum volume confidence sets for parameters of normal distributions. *AStA Advances in Statistical Analysis* 101: 309–320.

Index

M-statistics: Optimal Statistical Inference for a Small Sample, First Edition. Eugene Demidenko.
© 2023 John Wiley & Sons, Inc. Published 2023 by John Wiley & Sons, Inc.

Printed and bound by CPI Group (UK) Ltd, Croydon, CR0 4YY

16/04/2025

14658370-0004